Biosocial Interplay During Elementary School

Pol A. C. van Lier • Kirby Deater-Deckard
Editors

Biosocial Interplay During Elementary School

Pathways Toward Maladaptation in Young Children

Springer

Editors
Pol A. C. van Lier
Clinical Developmental Psychology
Vrije Universiteit Amsterdam
Amsterdam, The Netherlands

Kirby Deater-Deckard
Department of Psychological & Brain Sciences
University of Massachusetts Amherst
Amherst, MA, USA

ISBN 978-3-031-07108-9 ISBN 978-3-031-07109-6 (eBook)
https://doi.org/10.1007/978-3-031-07109-6

This Springer imprint is published by the registered company Springer Nature Switzerland AG
The registered company address is: Gewerbestrasse 11, 6330 Cham, Switzerland

Contents

Prologue: Introduction

Kirby Deater-Deckard and Pol A. C. van Lier

We are pleased to present to you the scholarship from some of the leading experts in research on social relationships in school during the primary/elementary school years. It is important to bear in mind when reading these chapters that all of the theory and empirical evidence being summarized (and most of the writing that was done) represented our lives prior to the COVID-19 pandemic. The seismic shift in our world due to the pandemic has had and will continue to have short-term and long-lasting effects on children's social relationships, school environments, and the underlying neurobiological processes that connect those experiences to adaptive and maladaptive development. The rapid adoption of home-based schooling via the Internet, and the corresponding social isolation of children, teachers, and parents, has disrupted learning and social environments and relationships in ways that we do not yet understand. Although we expect that school-based contexts and social relationships will gradually shift back to how things operated prior to the pandemic, there also are likely to be lasting changes in the locations and modes of communication that are used in elementary and secondary education settings. These lasting changes will have far-reaching implications for how we study and understand the powerful influences of children's relationships with their peers and teachers "at school," even if for some children those experiences are mediated almost entirely through digital devices.

Our collection of chapters addresses children's experiences in school-based peer and teacher relationships, with an emphasis on links with concurrent and subsequent neurobiological changes that may help explain the development of

K. Deater-Deckard (✉)
Department of Psychological & Brain Sciences, University of Massachusetts Amherst, Amherst, MA, USA
e-mail: kdeaterdeck@umass.edu

P. A. C. van Lier
Clinical Developmental Psychology, Vrije Universiteit Amsterdam, Amsterdam, The Netherlands

P. A. C. van Lier, K. Deater-Deckard (eds.), *Biosocial Interplay During Elementary School*, https://doi.org/10.1007/978-3-031-07109-6_1

behavioral, emotional, and academic problems. After the transition to primary education, children have to function in the context of age-matched peers, supervised by teachers. From 10% to 20% of children end up being poorly accepted, rejected, or outright victimized by their peers, receive low levels of support and potentially even conflictual relations with teachers, and end up perceiving the classroom and school on the whole as being an environment in which they have little support or enjoyment. Decades of research have shown the detrimental effects of such negative social experiences at school.

The current book provides an overview of how such experiences affect children's neurobiological factors, to better understand why these children develop such maladaptive outcomes. Chapters "Prologue: Introduction" and "Elementary School Social Experiences with Peers and Teachers: Manifestation and Development" present the state of the art in theory, methods, and conclusions from research examining developmentally typical and maladaptive peer and teacher relationships in schools and their links with children's social-emotional, cognitive, and academic functioning. These chapters provide an essential overview of the manifestations of social relationships in school settings, their complexity, interrelations, and pathways leading to healthy and maladaptive outcomes.

Then, with these foundations in mind, chapters "How Peers and Teachers Shape Elementary School Children's Academic and Socioemotional Development" and "School Social Relations and Child Development: Gene-Environment Interplay" integrate that knowledge regarding school environments with genetic factors that are correlated and interacting with environmental factors, to contribute to developmental changes in neurobiology, cognition, emotion, and behavior. This includes consideration of how children's genetically influenced attributes can elicit particular social experiences and make some children more susceptible to the impacts of these powerful social environments at school. In addition, work is presented summarizing how experiences can modify DNA via epigenetic mechanisms, in ways that create lasting changes in gene expression that alter underlying neurobiological functions in ways that influence developmental outcomes.

Following this, chapters "The Impact of School Social Experiences on Socioemotional and Behavioral Problems: The Hypothesized Role of DNA Methylation" and "Biological Embedding of Peer Experiences: The Contribution of Peer Adversity to Stress Regulation" present complementary literatures regarding development of self-regulation with particular emphasis on the acute and chronic stressors that arise in school social contexts that include rejection and victimization. Chapter "The Impact of School Social Experiences on Socioemotional and Behavioral Problems: The Hypothesized Role of DNA Methylation" examines the ways in which these early negative social experiences at school can lead to neurobiological changes in stress reactivity and self-regulation via the autonomic nervous system – changes that can have adaptive and maladaptive consequences later. Chapter "Biological Embedding of Peer Experiences: The Contribution of Peer Adversity to Stress Regulation" builds on this, focusing on theory and empirical evidence regarding how such experiences and neurobiological shifts can alter children's affective and social decision-making.

Finally, chapter "School Social Relations, Self-Regulation, and Social Decision-Making" addresses theory and current evidence regarding children's social relationships and their potential impacts on neural functioning. This chapter summarizes the literature using functional EEG and MRI, to more directly examine the correlated and interacting gene-environment mechanisms that are operating through stress reactivity and self-regulation processes, observed indirectly as changes in neural activity. These neural activity changes ultimately will help us understand how neurobiological alterations influence positive and negative developmental outcomes. We then complete the book with a concise Commentary to summarize and provide an initial integration of the key points across the chapters.

Collectively, the chapters provide an overview on the associations between adverse social experiences and child maladaptive outcomes. Importantly, they provide an overview of crucial pathways through which adverse social experiences in everyday elementary school settings may cause lasting changes in children's underlying neurobiology – developmental adaptations to social adversity that will help explain the well-established link with subsequent developmental psychopathological outcomes.

Elementary School Social Experiences with Peers and Teachers: Manifestation and Development

Antonius H. N. Cillessen, Nathalie A. H. Hoekstra, and Hannah K. Peetz

An important part of children's social experiences takes place in the context of school. In elementary school, young children spend their time primarily in a classroom context with a relatively stable group of classmates. The size and composition of the classroom can vary considerably depending on a range of circumstances. There are variations around the world in the time and duration of children's elementary school experiences. In the United States, for example, children begin to attend elementary school when they are 6 (grade 1) for either 5 years, after which they transition to middle school, or 6 years, after which they transition to junior high school. In the Netherlands, children are required by law to begin elementary school when they turn 4 years old (group 1) and are in elementary school for 8 years (groups 1–8), after which they transition to a system of secondary education with three main tracks (lasting 4–6 years). In other countries and cultures, arrangements are comparable or vary in details. What is common is that for a substantial number of years in the first decade of their lives, young children spend a considerable number of hours each day in the presence of classroom peers, guided by teachers. This chapter focuses on these two important socializing agents of influence on children's development in the classroom context, that is, their peers and teachers.

The starting point of our literature review was what in each country is understood by "elementary school." As indicated, what is meant by elementary school varies somewhat from country to country. Thus, our literature review was not guided strictly by age, but by the grades determined by the boundaries of the elementary school context and the classroom contexts it includes. As indicated above, this corresponded with grades 1–5 or 1–6 (ages 6–11 years or 6–12 years) in North

A. H. N. Cillessen (✉) · N. A. H. Hoekstra · H. K. Peetz
Behavioural Science Institute, Radboud University, Nijmegen, The Netherlands
e-mail: toon.cillessen@ru.nl

P. A. C. van Lier, K. Deater-Deckard (eds.), *Biosocial Interplay During Elementary School*, https://doi.org/10.1007/978-3-031-07109-6_2

American publications, "groups" 1–8 (ages 4–12) in Dutch school settings and publications, and in between arrangements for data from other countries.

Chapter Overview

This chapter addresses children's normative development. We focus on children's normative social and academic development in school, as well as their well-being. We do not address the occurrence of serious forms of developmental psychopathology. To provide a conceptual framework, we used a matrix for the discussion of the socializing influences of peers and teachers in elementary school, presented in Table 1. On the "predictor" side, we distinguished four levels at which the socializing influence of peers and teachers can be examined. This was based on a common distinction made in research on peer relationships (see Bukowski et al., 2018), and we extended this to the socializing influence of teachers. On the "outcome" side, we distinguished three domains.

We discussed the socializing influence of peers in elementary school at four levels: the level of children's individual characteristics (e.g., acceptance, rejection, popularity), the level of interactions or social-interactive behaviors (e.g., aggression, conflict, prosocial behavior), the level of dyadic relationships (e.g., friendships, mutual dislike), and the levels of groups and group processes (e.g., bullying, victimization).

We extended this distinction to the socializing influence of teachers. Here too we distinguished the socializing influence of teachers in terms of individual characteristics of teachers (e.g., work experience, classroom management style), the level of teacher-student interactions (e.g., expectations, attention, giving compliments), the level of teacher-student relationships (e.g., support), and again the level of group processes in the classroom (bullying, victimization, classroom norms).

Table 1 Matrix for the socializing influence of peers and teachers in elementary school

	Outcomes		
Predictors	Academic functioning	Social functioning	Well-being
Peers			
Individual characteristics			
Interactions			
Relationships			
Group processes			
Teachers			
Individual characteristics			
Interactions			
Relationships			
Group processes			

On the "outcome" side, we distinguished three broad categories of outcomes in children's development: academic functioning, social functioning, and well-being (social-emotional functioning). Each of these three is an important dimension of children's functioning in schools and often represented as such in measures of children's classroom functioning (see, e.g., Hightower et al., 1986).

Together, this resulted in a matrix of four levels of peer influence on three outcome domains and a similar matrix of four levels of teacher influence on the same three domains of child outcomes. In the two main sections below, we describe relevant knowledge and findings for each element of each matrix. We also address recommendations for further research. We close this chapter with a conclusion.

The Socializing Influence of Peers in Elementary School

This section focuses on the manifestation and the different layers of the peer system, namely, individual characteristics, social interactions, dyadic relationships, and group processes, and their impact on children normative academic, social, and emotional development (i.e., well-being) in elementary school.

Individual Characteristics

Definition At the level of individual characteristic, we focus on two domains. The first is social status (Cillessen, 2009; Cillessen & Marks, 2011). The social status of children in elementary classroom peer groups is usually derived from peer evaluations of liking and disliking and results in continuous scores for social status and a classification in sociometric status types. The continuous dimensions are acceptance (liking received), rejection (disliking received), social preference (difference between acceptance and rejection), and social impact (sum of acceptance and rejection). The traditional system of sociometric status groups uses these continuous scores to classify children as accepted, rejected, controversial, neglected, or average in sociometric status. Social status can be based on (dis)liking but also on direct peer evaluations of (un)popularity. A continuous score for popularity often is derived from peer nominations of "who is most popular" and "who is least popular." This score can be used to classify children in popularity groups, but this is less commonly done. Peer acceptance and preference are often referred to as sociometric popularity, whereas popularity per se is often referred to as perceived popularity (Parkhurst & Hopmeyer, 1998).

Children's individual characteristics also include their social cognitions. Social cognition is used here as a broad term referring to how children think about themselves and others. Children's social cognitions are based on their developmental history of interactions with adults and peers. According to models of children's social cognitions related to peer relations, children build a database of cognitions and expectations in interactions with others (e.g., Crick & Dodge, 1994; Dodge, 1986). Children use this database to make decisions about how to behave in social situations and how to interpret the behaviors of peers. Two well-known phenomena are children's hostile attribution biases and their role in aggressive behavior (Crick & Dodge, 1994).

Academic Functioning The connection between children's functioning with their peers in school and their academic functioning is an important topic to address. The social processes of children in interactions with their peers can either facilitate or impair their learning. This is clear from a large body of research on the associations between sociometric status (based on peer acceptance) and academic outcomes. Newcomb et al. (1993) conducted an extensive meta-analysis of the research findings on the associations between the traditional sociometric status groups (accepted, rejected, controversial, neglected, and average) and measures of academic functioning in school, combined in a broadband category labeled "cognitive abilities." The meta-analysis showed that peer acceptance is positively associated with academic functioning, whereas peer rejection is negatively associated with academic functioning. That is, children who have positive relationships with peers in the classroom do better in school, whereas children who have difficulties in their interactions with peers in the classroom do less well in school. The association between peer acceptance and academic achievement is expected to be reciprocal (Hughes & Chen, 2011; Hughes et al., 2014). On the one hand, not being able to work well with others impairs (collaborative) learning. Children who are disliked or rejected may be distracted by their worries about these negative interactions and therefore not able to direct their attention to learning. On the other hand, children who do less well than others academically may be perceived negatively by their classmates because they deviate from the norm, do not meet teacher expectations, or disrupt classroom processes when they are behind. Underlying third variables may also be at play. For example, (social) cognitive deficits may explain both why children do less well with their peers and less well on their school work. This hypothesis needs to be examined further in future research.

The second type of social status is (perceived) popularity. The empirical database for its association with academic achievement is smaller so far. Furthermore, this issue has been considered more among (early) adolescents in middle school, junior high school, or the beginning years of secondary education and less frequently in elementary school. This is partly because in the younger grades of elementary school, "popularity" is not yet a central construct, but begins to emerge in the second half of elementary school. Previous findings on the associations between popularity and academic functioning were mixed (see Schwartz et al., 2006). This has led researchers to hypothesize that the association between popularity and academic functioning is moderated by other variables. One moderator is aggression. Schwartz et al. (2006) demonstrated that children who are popular and aggressive score lower on academic functioning, whereas children who are popular but not aggressive do as well as others. Thus, the "popular-aggressive" group is a concern in terms of academic functioning (see also Rodkin et al., 2000).

Social Functioning When examining the effects of children's individual peer-related characteristics on their social functioning, an important individual factor is formed by children's social cognitions. Here we focus specifically on cognitive biases or tendencies such as the hostile attribution bias. The attributions that children make for their social experiences are related to their social functioning. Specifically, the attributions that children make for their negative experiences have been related to their perceived social support and acceptance in the peer group (Visconti et al., 2013). Children who attribute peers' negative behavior to their jealousy report feeling more accepted than children who blame themselves for their peers' negative behavior. Whereas this study showed that children's attributions are related to their perceived acceptance in the peer group, other studies have shown that they are also related to their actual acceptance. For example, children's hostile attributions have been related to subsequent peer rejection. Specifically, there seems to be a reinforcing loop between peer rejection, hostile attributions, and aggression, highlighting the pivotal role of cognitions in children's behavior and social functioning (Lansford et al., 2010). These examples indicate how children's cognitions are related to their social functioning at school, indicating that children might feel less accepted depending on their cognitions but also that they might actually suffer social consequences related to their cognitive biases.

Well-Being To assess the effects of peer relations on children's well-being, researchers have examined loneliness, depression, (social) anxiety, and the experience of stress. Across sources of information, the absence of peer acceptance and low popularity are consistently associated with loneliness (Geukens et al., 2021). Whereas loneliness can be seen as a normative developmental outcome, depression and social anxiety are related but in the clinical domain of internalizing problems. Here too, there are consistent associations of low status (preference or popularity) with depression and anxiety (Prinstein et al., 2018). Sandstrom and Cillessen (2003) used a daily diary method to assess children's feelings in the classroom related to their peer experiences. They also demonstrated that rejection experiences are associated with reported stress and less well-being. An important question is whether children would seek help for their social difficulties of children in the classroom. Recently, whether or not children would report their social difficulties in the classroom was investigated in relation to being bullied (van der Ploeg et al., 2021).

Children's cognitions also play a role in their well-being. The study by Visconti et al. (2013) mentioned above also showed that the attributions children made for their peers' negative behavior was related to their feelings of loneliness. When children blamed themselves, they reported greater loneliness. Furthermore, cognitions are not only concurrently related to well-being but also prospectively. Children's self-blaming as well as hostile attributions were related to internalizing and externalizing problems respectively later on (Perren et al., 2013). It appears that children experience lower well-being when they blame themselves for their peers' negative behaviors. Thus, breaking these maladaptive attributional tendencies is a fruitful avenue for interventions aimed at increasing children's well-being. Such interventions have been shown to be effective for adolescents (Yeager et al., 2014) and might similarly be effective for children.

Interactions

Definition Interactions or social-interactive behaviors refer to the one-on-one interactions between peers. It refers to the specific behavior of one child against another in a dyad or a group. The most commonly studied interactive behavior in the peer relationships literature is aggression, but other behaviors are studied as well. Typically, interactive behaviors are divided in three main groups: aggressive and antisocial behavior, prosocial behavior, and social withdrawal (or the absence of interaction) (McDonald & Asher, 2018). Social behavior can be measured with peer nominations or teacher ratings, but the ideal method is behavior observations that allow for detailed analysis of streams of interactive behavior. Specific analysis methods have been developed that are well suited for the analysis of social interactive behavior in dyads or groups, such as the Actor-Partner Interdependence Model (e.g., Kenny et al., 2006). Children's peer interactions and social-interactive behaviors are often studied in the context of specific situations or settings, such as the occurrence of conflict, or paradigms, such as unstructured playgroups, cooperative and competitive tasks, and the peer group entry paradigm.

Academic Functioning Children's peer interactions have been shown to be associated with their academic functioning. For example, Gommans et al. (2015) examined collaboration quality and knowledge gain of children in dyads during a collaborative learning task on a computer. Mutual listening had a positive effect on knowledge gain, whereas dominance had a negative effect. This was further qualified by the role of the children in the dyad, as one was more popular and the other less popular. Specifically, the knowledge level of the more popular dyad member predicted the knowledge gain of the less popular member later, specifically when there was more mutual listening and less dominance of the more popular member. Thus, the way in which children interact with classmates during moments of collaborative learning may be indicative for their academic outcomes.

The interactions between children in a group (rather than a dyad) have also been shown to predict academic functioning of the group as a whole. One consistent finding in the literature is that children who are aggressive or disruptive in the classroom (as seen by peers or teachers) also score systematically lower on measures of academic performance, either as seen by teachers or on objective tests (Cillessen & van den Berg, 2012). Conversely, peer acceptance and prosocial behaviors typically correlate positively with measures of academic functioning (Newcomb et al., 1993).

Social Functioning Children's social interactions and social-interactive behaviors with peers are closely related to their social functioning. Coie et al. (1990) conducted a narrative review of the literature on the impact of children's social behavior on their peers. Their review included teacher ratings, peer reports, and systematic observations of social behaviors and interactions. Several observational studies used a clever design in which children interacted in unfamiliar peer groups – they had no previous reputation with each other (e.g., Coie & Kupersmidt, 1983; Dodge, 1983). This research clearly shows that children are primarily "data based" in their perceptions of their peers. Aggressive interactions lead to disliking, whereas prosocial interactions lead to liking.

Longitudinal studies have shown that the associations between children's social-interactive behaviors and their peer relations are reciprocal (e.g., Ladd, 2006). For example, whereas aggression leads to peer dislike and rejection, rejection may lead to further aggression, because rejection is frustrating or because rejected children learn that this is a way to get others' attention or to get what they want. This leads to a vicious circle, or, more formally, cascading effects across development, in which "the rich get richer" and "the poor get poorer." Prosocial behaviors lead to positive interactions with peers, which then form a training ground in which children can further improve and develop their social skills and competence. Aggression and antisocial behaviors lead to exclusion and rejection, which deprives children of the opportunity to improve their behavior and social skills. This further emphasizes why effective interventions are important to change a downward negative spiral into a positive direction.

Well-Being What are the effects of children's social interactions and social-interactive behaviors with peers on their well-being? The interactions that children have with their peers also impact their well-being. Especially the absence of interactions can have severe consequences. For example, the absence of social interactions during playtimes has been shown to be associated with loneliness, depressive symptoms, and social anxiety (Coplan et al., 2015). Similar results have consistently been observed with regard to social exclusion. Children that are being ostracized in group play experience a lower psychological well-being and higher levels of negative mood after not being able to take part in social interactions (Lansu et al., 2017). Specifically, having even one interaction partner reduces the negative effects of ostracism substantially (Sandstrom et al., 2017), indicating the importance of social interactions for children's well-being. It is particularly clear that negative interactions with peers have a negative impact on children's well-being at school. Children who are excluded from interactions with their peers, who are rejected or ostracized in the peer group, or who are the targets of gossiping and exclusion develop externalizing behavior problems or internalizing behavior problems as a result (Asher & Coie, 1990).

Dyadic Relationships

Definition What do we mean by dyadic relationships? This addresses the type of relationship between two children. Dyadic relationships are separate from individual characteristics such as status and social-interactive behaviors such as aggression. Dyadic peer relationships are often formed at school where children spend much time each day and play an important role in their academic and social development and well-being. The most commonly studied dyadic relationship is friendship. There is a large literature on children's friendships (see, e.g., Bukowski et al., 2018). Other types of dyadic relationship are important as well, such as relationships based on mutual dislike, mutually aggressive dyads, bully-victim pairs, or bully-defender relationships. These relationships are less commonly studied. In adolescence, romantic relationships are important, but they are typically not studied in the elementary school context.

Academic Functioning What are the effects of children's dyadic relationships with peers on their academic functioning? The benefits of friendships, in terms of positive social functioning and well-being, may mediate an association with better school performance. For example, Zucchetti et al. (2015) examined the reciprocal associations between friendship and achievement over time and found that positive friendship quality in the third grade predicted school achievement in the fourth grade. Other research has shown that friends affect each other's academic functioning. Gremmen et al. (2018) showed that students' own engagement and achievement were associated with their friends' engagement and achievement. Longitudinal social network analyses indicated that students increased in school engagement and achievement when their friends did. This shows that considering children's friendship dyads is important for understanding their academic functioning at school. On the negative side, children who are involved in difficult relationships, such as being in a bully-victim dyad or a mutually aggressive pair, are expected to do less well in terms of their academic performance. Here it is difficult to discern whether this would be due to their specific dyadic relationship or to the general tendencies to be aggressive or the target of others' aggression.

Social Functioning What are the effects of children's dyadic relationships with peers on their social functioning? There is a large literature on the social behavior of children in friendship dyads. Much of this research has been conducted using careful and detailed behavior observations. Children in friendship dyads have been observed in interactions with one another and compared to interactions between non-friends. This research in general shows that children in friendship dyads score generally more positively on prosocial behaviors, sharing and helping, than children in non-friend dyads (Hartup, 1996). However, Hartup et al. (1988) also demonstrated that children in friendship dyads have more conflict than children in non-friend dyads. This points to an important function of friendship – friendships are not only for mutual support but also a context that fosters the learning of critical social skills, such as conflict resolution. Finally, not all friendships are alike. There are also much variations within friendship dyads, related to other characteristics of the children in the dyad. Peters et al. (2010) examined variations between friendship dyads in social status and how this is related to variation in social behaviors. Thus, not all friendships are alike, but the studies on children's social functioning in friendship dyads show that friendships are a context for positive behaviors and mutual support but also a context in which to learn critical social skills, such as conflict resolution (Hartup, 1996).

There are fewer observational studies on children's social interactions in other types of dyads. One exception is an observational study by Coie et al. (1999) who observed boys in mutually aggressive dyads in playgroups of six boys. They found that mutually aggressive dyads displayed twice as much aggression as randomly selected dyads and attributed greater hostile intentions toward each other, which may explain their greater aggression toward each other. Hubbard et al. (2001) further examined the social cognitions related to dyadic aggression and found that hostile attribution biases toward a particular peer were related to reactive aggression toward that peer. Given these findings, a question is whether children's social functioning in dyads can be improved to reduce hostility in the classroom. In an intervention study, van den Berg et al. (2012) used seating arrangements to improve children's dyadic relationships and interactions in the classroom. (This study is also discussed below.) Given the importance of children's dyadic relationships for their learning and social behavior, this effort is important.

Well-Being Friendships are critical for children's well-being. One important provision of friendship is social support, making it a buffer for negative experiences (see, e.g., Bagwell & Bukowski, 2018). Indeed, friendship quality serves as a moderator between victimization experiences and well-being (e.g., Cuadros & Berger, 2016). Friendships, especially when of good quality, alleviate the stress of negative peer experiences (Peters et al., 2011). In addition, friendship is a context in which children practice their social skills. Given the many provisions of friendship, it is not surprising that having good-quality (friendship) relationships is positively associated with children's well-being (Bagwell & Bukowski, 2018). Conversely, being involved in enmity, an aggressive dyad, or a bully-victim pair does not bode well for children's well-being, although here again it is hard to distinguish the relationship effects from general tendencies to be aggressive or victimized. This has been addressed in some studies (see, e.g., Peters et al., 2010, on friendship as a buffer for victimization). The effects of positive dyadic relationships on children's well-being at school further emphasize the importance of interventions aimed at improving children's dyadic peer relationships in the classroom as well as overall classroom climate.

Groups

Definition Group processes among peers include group dynamics and behavior and group norms. First, group processes refer to the behaviors that involve multiple group members at once. One important group process is bullying. Although bullying has been described in the past as a dyadic process involving a bully and a victim, it is now considered a group process in which the whole peer group is involved in some way (Salmivalli et al., 1996). Bullying can be defined as the repeated harassment of the victim by one or more bullies, in which there is a power imbalance between the victim and the bully (Olweus, 1994). Other peers are often involved in this process either by defending the victim, by following or reinforcing the bully, or by trying to stay out of it completely (Salmivalli et al., 1996). The group effects of bullying are markedly demonstrated by the healthy context paradox, to which we refer further below. A second important focus at the level of groups is classroom norms. Classroom norms dictate which behaviors are accepted in the group; we discuss them briefly within the context of bullying and peer victimization. Below we address how children's experiences with bullying, and the related group norms, impact their academic functioning, social functioning, and well-being.

Academic Functioning Peer victimization has been shown to be related to school avoidance and lower levels of classroom engagement (Buhs et al., 2006). This association depends on the type of victimization, with physical harassment being related to school avoidance and relational harassment to lower engagement. Because both school avoidance and engagement are important contributors to academic achievement, peer victimization is a risk factor for lower academic attainment.

Social Functioning Research has shown that group processes also play a role in children's social functioning. For example, children's role in a group process such as bullying is associated with their status in the group. Research examined how defending a victim would be rewarded by the peer group and found that defenders were both liked more and perceived as more popular by the victim they defended (Sainio et al., 2011) and were also rewarded with higher popularity in the peer group as a whole (van der Ploeg et al., 2017). Furthermore, peer victimization is bidirectionally related to social functioning. Research has indicated that children who have lower levels of social skills or who have a lower social status are at a higher risk for being victimized (e.g., Sentse et al., 2015). Additionally, being victimized was related to subsequent lower social status.

On the side of the bully, bullying is often related to higher social status in the form of popularity, although bullies are often disliked as well (e.g., Pouwels et al., 2018). Classroom norms, specifically popularity norms, play a role in how accepted bullying behavior is in the classroom. When popular children engage in bullying, the association between bullying and peer rejection becomes weaker (Dijkstra et al., 2008). This indicates that the behavior of popular children impacts how the group perceives bullying behavior.

Well-Being The impact of peer victimization on well-being has been extensively studied. Several meta-analyses highlight the negative consequences of victimization (e.g., Gini & Pozzoli, 2009; Moore & Woodcock, 2017; Reijntjes et al., 2010). For example, victimization is related to mental health problems, such as depression, social anxiety, and suicide intentions. Victimization is also related to externalizing problems such as aggression and risk taking. Interestingly, low well-being has also been identified as a precursor to victimization. Children who experience internalizing problems, such as depression, are more likely to be victimized (Forbes et al., 2019).

The negative consequences of victimization appear to be even stronger in a so-called healthy context. In an environment in which there are few victims, the victims experience higher levels of internalizing problems than victims in an environment with several other victims (Huitsing et al., 2012). This especially poses a problem for interventions to reduce bullying, as the remaining victims show decreases in well-being (Huitsing et al., 2019).

The Socializing Influence of Teachers in Elementary School

This section focuses on the socializing influences of teachers on children's normative academic, social, and emotional development (i.e., well-being) in elementary school. Teachers play a large role in children's development, directly and indirectly, in the way they manage the classroom peer group. In this section we examine individual characteristics of teachers, student-teacher interactions and relationships, and teachers' role in classroom group processes (especially focusing on bullying and group norms).

Individual Characteristics

Definition Teachers bring with themselves a number of individual characteristics into the classroom. These include their previous training and experience, their self-efficacy expectations regarding their own effectiveness as a teacher and abilities to manage the classroom, and any expectations or possible biases they may have regarding their students.

Academic Functioning The individual characteristics that teachers bring into the classroom are likely to affect their teaching. Hence, teachers' individual characteristics may be related to students' academic functioning. The central construct in the link between teachers' characteristics and students' academic outcomes seems to be teacher quality (e.g., Croninger et al., 2007; Darling-Hammond, 2000; Gerritsen et al., 2017; Harris & Sass, 2011; Stronge et al., 2007). Not only objective, observable variables such as experience, training, or professional development are studied within the scope of teacher quality research, but the student perspective is important as well. Fauth et al. (2014) found that student ratings of classroom management were positively associated with student achievement and that ratings of cognitive activation and classroom climate were positively related to students' subject-related interest. In general, research on teacher quality has indicated that teacher experience plays an important role. In their study of Dutch twins who were assigned to different classrooms, Gerritsen et al. (2017) found that twins who had a more experienced teacher did better in both reading and mathematics. However, Huang and Moon (2009) did not find that teachers' total years of experience predicted student achievement, but their years of teaching experience in a specific grade was related to reading performance. Teacher experience is closely related to teachers' productivity, according to Harris and Sass (2011). Teacher productivity is defined as the extent to which teachers contribute to students' achievement, and has been shown to increase with experience, most rapidly in the first few years of being in the profession (Harris & Sass, 2011). Stronge et al. (2007) found that highly effective teachers are specifically stronger than less effective teachers in instruction, student assessment, classroom management, and personal qualities.

Social Functioning Teachers have an impact on children's social competence development and their social behavior with their peers. Teachers are a role model of social behavior and serve as a social referent for how to interact with others (Hughes et al., 2001). When a teacher is more supportive of students, they show more prosocial behavior and more positive peer relationships with classmates (Hendrickx et al., 2016). Conversely, for children who are rejected by their peers, the teacher can be an important factor in reducing the chance of additional peer group difficulties (Elledge et al., 2016) or exacerbating the social difficulties of rejected children (Lucas-Molina et al., 2015). Teachers can play an important role in promoting the development and growth of social skills. However, negative interactions with teachers can contribute to maladjustment and behavioral problems (Brendgen et al., 2006).

Well-Being Teachers' individual characteristics have also been related to children's well-being at school. Oldenburg et al. (2015) investigated how teacher characteristics were related to peer victimization in elementary school classrooms. They found higher victimization rates in classrooms of teachers who attributed bullying to external factors and in classrooms of teachers who had a personal history of bullying peers themselves. This shows that teachers' individual experiences may indirectly influence students' well-being through the way they handle classroom processes such as bullying. Similarly, teachers' individual characteristics have been shown to predict how they respond to threats to students' well-being, such as victimization. Troop-Gordon and Ladd (2015) found that teachers who view victimization as a normative process less often reprimand harassers and are more likely to suggest to victims that they avoid the harasser or solve the problem by themselves.

Interactions

Definition The interactions that take place in the classroom affect students' school lives. On the one hand, there are the actual interactions between teacher and students. These interactions are likely to reflect teachers' personal perspectives, such as their academic expectations. These refer to the extent to which teachers expect their students to do well in school (e.g., McKown & Weinstein, 2008; Rubie-Davies et al., 2006) or their feelings of self-efficacy as a teacher, that is, their beliefs in their own ability to effectively fulfill their role as a teacher (e.g., Caprara et al., 2006; Skaalvik & Skaalvik, 2007). On the other hand, teachers shape the interactions among their students. Studies focusing on teachers' contributions to classmates' peer relationships and their interpersonal development at school refer to this phenomenon as "the invisible hand of the teacher" (Farmer et al., 2011). According to Farmer and colleagues, teachers have a twofold role when it comes to shaping youth's social development, as they are both an authority regarding rules and expectations for social behavior and a facilitator of social interactions among their students.

Academic Functioning Research has shown that students are able to distinguish between high and low teacher expectations based on their teacher's verbal and nonverbal behaviors (Babad, 1990). Hence, the interactions between students and teachers affect students' academic functioning through teacher expectations. Teachers can have expectations of individual students but also of the classroom as a whole. With regard to the individual level, the widely known study by Rosenthal and Jacobson (1968) showed how so-called Pygmalion effects can affect student achievement outcomes. Teachers were told at the beginning of the year that for some of their students, who were in fact chosen at random, a growth spurt with regard to academic achievement could be expected. The increased expectations that teachers had of these students then were associated with larger gains in intellectual development for these students compared to regular students, especially in the lower grades of elementary school. A recent study regarding trajectories of teachers' academic expectations of individual students found that student achievement may not only be an outcome in this mechanism but also a moderator. Wang et al. (2020) found that trajectories of teacher expectations throughout the year differed between high- and low-achieving students in the sense that high-achieving students were systematically overestimated, whereas low-achieving students were systematically underestimated. Similarly, teachers' class-level expectations were also related to students' academic functioning, as their expectations of the class as a whole guide their interactions with students (Rubie-Davies, 2007). In Rubie-Davies' study, teachers were observed twice in the school year during reading lessons. The study found differences in teachers with high, average, and low expectations in their interactions with their students. Specifically, teachers with high expectations of their classroom provided students with more instruction and explanation about content, gave more feedback, posed more higher-order questions, and used more positive behavior management strategies.

Social Functioning The invisible hand of the teacher has predominantly been studied in relation to the teacher's role in shaping peer relationships. Studies indicate that teachers mostly function as a social referent through their behavior. How a teacher interacts with a student is directly related to the acceptance of that student in the peer group and has shown to be a functional mechanism between the child's behavior and their acceptance in the group (McAuliffe et al., 2009). Especially negative behavior by the teacher towards a student shapes classmates' dislike towards that student (Hendrickx et al., 2017). Conflict between a teacher and a student thus impacts not only their relationship but also how the student is viewed by the entire class.

Well-Being Whereas the effects of teacher expectations on students' academic outcomes have been studied repeatedly, less is known about the potential effects on socio-emotional outcomes. Rubie-Davies et al. (2020) recently demonstrated the effects of class-level teacher expectations on students' beliefs at the end of the school year. They found that the beliefs of students with a high-expectation teacher were higher by the end of the year than those of students with a low-expectation teacher. For example, when students perceived their teacher to have higher expectations, they also perceived their teacher as more supportive, which is expected to contribute to their well-being at school.

Dyadic Relationships

Definition The everyday interactions of teachers with their students, as well as a complex interplay of beliefs, attitudes, and behaviors, form a relationship between a teacher and individual students over the course of the school year (Hamre & Pianta, 2006). Teacher-student relationships (TSRs) have been widely studied, and their importance has been emphasized repeatedly (e.g., Hamre & Pianta 2001, 2006; Sabol & Pianta, 2012; Verschueren & Koomen, 2012; Wubbels et al., 2006, 2014). In research focusing on TSRs, different theoretical approaches are used, among which a framework based on attachment theory and an interpersonal theory framework. In the attachment-based framework, scholars consider TSRs in terms of three dimensions: closeness, conflict, and dependency (Koomen et al., 2012; Pianta, 2001). A TSR characterized by a high level of closeness and low levels of conflict and dependency is considered a safe haven and a secure base from which students can explore the world (Verschueren & Koomen, 2012). Teacher-rated closeness and especially conflict were moderately stable from kindergarten through the sixth grade (Bosman et al., 2018; Jerome et al., 2009; Pianta & Stuhlman, 2004). Jerome et al. (2009) also found that boys had lower closeness and higher conflict scores than girls and that this gap increased in the mid-elementary school years. In interpersonal theory, TSRs are studied in terms of the dimensions of agency and communion in the interpersonal circumplex (Wubbels et al., 2006). All interactions in a TSR can thus be seen as a combination of these two dimensions. High levels of teacher agency and communion have been associated with effective TSRs (e.g., Wubbels et al., 2006). In addition to these two main approaches, other theoretical frameworks have been used in TSR research. Different studies adopt different approaches, but all have in common that they investigated the associations between TSRs and a wide array of socio-emotional and academic outcomes. Whereas research was first mainly focused on the teacher's perspective, more and more studies of TSRs now include the students' perception as well (e.g., Brinkworth et al., 2018; Koomen & Jellesma, 2015).

Academic Functioning One of the most prominent interests in TSR research are students' academic outcomes, as teachers have the potential to promote students' academic functioning through the formation of positive relationships. Within the scope of research on TSRs and academic functioning, studies have focused on outcomes such as students' achievement and motivation/engagement (e.g., Furrer et al., 2014; Sabol & Pianta, 2012. The link between TSRs and student achievement is well established and has been proven repeatedly. In their widely known study, Hamre and Pianta (2001) showed that these associations even last in the long run, as they found that negative aspects of kindergarten TSRs were related to academic achievement in upper elementary school. Similarly, the meta-analysis of Roorda et al. (2011) indicated that positive TSRs were positively related to both achievement and engagement and that negative TSRs were negatively related to both outcomes. They found that for elementary school students, TSRs were more strongly associated with engagement than with achievement and that engagement may mediate the association with achievement. An updated meta-analysis by the same authors indeed showed that engagement partially mediated the association between TSRs and student achievement (Roorda et al., 2017). In addition to studies investigating student achievement as affected by TSRs, other studies found differences in TSRs between students of different achievement levels. For example, Jerome et al. (2009) found that students who started in kindergarten with lower achievement levels had higher levels of teacher-perceived conflict, lower levels of teacher-perceived closeness, and lower-quality TSRs in general throughout the sixth grade.

Social Functioning The interactions between students and teachers not only affect the academic side of students' school lives but also the social side, as TSRs have been shown to have an effect on students' perceptions of one another. For example, De Laet et al. (2014) examined the reciprocal associations of the TSR with perceived and sociometric popularity. They found that sociometric popularity was positively associated with teacher-child support, which in turn further predicted sociometric popularity. A higher perceived popularity was related to more conflict in the TSR, which in turn further increased perceived popularity.

Well-Being The TSR has also been shown to be an important factor for students' well-being in the classroom. Pianta and Stuhlman (2004) obtained parent and teacher reports of internalizing and externalizing behavior of first-grade students and found associations with closeness and conflict in TSRs. Less closeness in the TSR, as reported by the teacher, was linked to higher mother-reported internalizing behavior. Higher levels of teacher-rated conflict in the TSR were positively related to mothers' reports of externalizing behavior. Comparable results were found for teacher-reported internalizing and externalizing behavior. Higher levels of conflict were associated with more internalizing and externalizing problems. Lower levels of closeness were related to more internalizing, but not more externalizing, behavior. Similarly, the meta-analysis by Lei et al. (2016) indicated that positive aspects of the TSR were negatively related to externalizing behavior problems and that the negative TSR components were positively associated with these problems. Lastly, although less focused on in TSR research, TSRs also seem to be associated with teachers' own well-being. For example, positive aspects of TSRs have been found to be associated with outcomes such as job satisfaction or work motivation, whereas the negative aspects seem to be linked with teacher stress and burnout (Spilt et al., 2011). Likewise, Aldrup et al. (2018) found that the link between student misbehavior and teacher well-being was mediated by the TSR. When students misbehave, teachers tend to report a lower TSR, and this was linked to a lower occupational well-being.

Groups

Definition In an elementary school classroom, several group processes are simultaneously at play. As it is beyond the scope of this chapter to provide a comprehensive overview, we focus on two group processes in which the teacher plays an important role: bullying and victimization and the classroom climate as a whole. Even though these are often studied as different constructs in the literature, they are naturally intertwined and are part of the larger, overarching process of the students forming a group together. As teachers are the leaders of their classrooms, they are in a key position when it comes to bullying and victimization, as they are the ones who can prevent the initiation or intervene in the continuation (Troop-Gordon, 2015). Both the extent to which teachers are aware of these processes (e.g., Ahn et al., 2013) and their beliefs and attitudes regarding them (e.g., Troop-Gordon & Ladd, 2015; Yoon & Kerber, 2003) may affect the group as a whole. Furthermore, two factors may be of importance in teachers' potential to create a positive classroom climate: teacher attunement and classroom seating arrangements. Teacher attunement refers to the extent to which the teacher's knowledge of the peer ecology is accurate and matches the students' perceptions (e.g., Hamm et al., 2011; Norwalk et al., 2016). Seating arrangements refer to both the physical layout of the room in terms of placement of desks and chairs (e.g., Wannarka & Ruhl, 2008) and to where individual students are seated (e.g., van den Berg et al., 2012).

Academic Functioning Both bullying and victimization and classroom or school climate in general have been linked to students' academic functioning. For example, Wang et al. (2014) found that victimization and school climate were both associated with a lower grade point average. Moreover, bullying and victimization and classroom or school climate are likely to affect one another, which is also shown by Yang et al. (2018). They found that victimization at the student level was negatively associated with student engagement and that this association was moderated by school climate. Likewise, the seating arrangement that teachers implement in their classroom has an effect on students' academic achievement. Hastings and Schwieso (1995) found that all students spent more time on-task during mathematics and English lessons when they were seated in rows rather than groups; especially the most disruptive students showed the largest increase in on-task behavior. Bennett and Blundell (1983) found that students produced more work in the same time when seated in rows, but the quality of their work remained the same as in groups. For other academic outcomes, not rows but seating arrangements in which students face each other showed better outcomes. Students in circles show a higher level of participation in class discussions (Rosenfield et al., 1985) compared to being seated in rows or groups, and the frequency of question-asking was higher when students sat in semicircles than when they sat in rows (Marx et al., 1999). Finally, Gremmen et al. (2018) showed that also the classmates next to whom students are seated may affect their academic achievement, as they found that the scores of near-seated peers diverged more over time if they were not friends.

Social Functioning The influence that teachers exert through these group processes has also been found to affect individual students' social functioning in the classroom and the functioning of the group as a whole. Veenstra et al. (2014) investigated how students' perceptions of their teachers' antibullying attitudes, efficacy, and efforts to reduce bullying were associated with bullying. They found that when students saw their teacher as efficient in decreasing bullying, there was less peer-reported bullying in the classroom. In addition, students' perceptions of their teacher's efforts to reduce bullying were related to a decrease in peer-reported bullying over time. Likewise, seating arrangements are important for students' social functioning. Van den Berg and Cillessen (2015) showed that the physical place where students sit in the classroom is associated with their likeability and popularity. Students who were seated more to the center of the classroom were liked better and perceived as more popular by their peers. Moreover, the study by van den Berg et al. (2012) even suggests that teachers could use seating arrangements as a tool to change peer affiliations. They found that decreasing the physical distance in the classroom seating arrangement between two students who disliked each other in the beginning of the school year led to increased likeability ratings, especially for students who were initially perceived most negatively.

Well-Being Finally, both of these group processes have been shown to contribute to students' well-being. Teachers' beliefs and attitudes toward bullying in particular seem to play an important part in student well-being. In order for teachers to support students who are victimized, they need to be able to identify them. However, Ahn et al. (2013) found that student-teacher agreement on bully-victim dyads was low (7.9%). Oldenburg et al. (2016) found this as well for teachers' identification of self-reported victims (approximately 25%). Norwalk et al. (2016) found similar levels of teacher attunement to victimization, ranging from 0 to 36% across schools, and investigated how it was associated with school belonging. They found that when teachers were more attuned to peer victimization in the Fall, their students had higher expectations of their peers to intervene in bullying situations in the Spring and reported a higher sense of school belonging.

Troop-Gordon and Ladd (2015) found that teachers' beliefs about peer victimization were associated with how they responded to students who reported being victimized. When teachers viewed peer victimization as a normative process, they were more likely to use passive strategies, such as advising the victim to avoid the bully, and they were less likely to punish the bully for their actions. This is in turn likely to affect victims' well-being, as they may think their teacher does not care or is not capable of doing anything about it (e.g., Yoon & Kerber, 2003). In addition to teachers' impact in bullying and victimization, the influence that they have on the group through seating arrangements has also been found to be related to students' well-being. Troop-Gordon and Ladd (2015) found that when teachers separated aggressors and victims, this was associated with less aggression over time and lower classroom levels of peer victimization. Similarly, van den Berg et al. (2012) demonstrated that classroom seating rearrangements can lead to less peer victimization and social withdrawal. Likewise, van den Berg and Stoltz (2018) found that students with externalizing behavior problems benefitted from sitting next to a prosocial, well-liked classmate. Over time, their teachers reported that their students showed fewer externalizing behavior problems and were better liked by their peers. Importantly, their seatmates did not show more aggression or less prosocial behavior. These studies suggest that students' well-being can be promoted through purposeful classroom seating arrangements.

Conclusion

An important part of children's socialization takes place at school. Children spend a substantial amount of time in a classroom context with a relatively stable peer group. We examined four levels of the socializing influence of peers and teachers and their impact on children's normative social development, academic development, and well-being. We examined the socializing influence of peers in elementary school at four levels: individual characteristics, interactions and social behaviors, dyadic relationships, and group processes. We also examined the socializing influence of teachers in terms of individual characteristics of teachers, teacher-student interactions, teacher-student relationships, and classroom group processes.

The impact of peers and teachers on children's development in school represents a large area of research. Our review covered a selection, based on our choice of predictors and outcomes. There are other relevant topics to discuss, which were beyond the score of this chapter. At the same time, there are important directions for future research. For example, the use of physical space and seating arrangements to improve the social dynamics of classrooms and schools is an important direction for further research. In addition, this chapter did not address the virtual space of online social interactions between students and between students and their teachers, but this is obviously a critical direction for further research as well. It will be particularly important to examine how online peer interactions and online teaching are similar to or different from the findings from the existing literature, primarily based on offline interactions. Together, this review demonstrates that the social context of peer and teachers continues to be a critical factor in elementary school children's academic development, social development, and well-being.

References

Ahn, H. J., Rodkin, P. C., & Gest, S. (2013). Teacher-student agreement on "bullies and kids they pick on" in elementary school classrooms: Gender and grade differences. *Theory into Practice, 52*(4), 257–263. https://doi.org/10.1080/00405841.2013.829728

Aldrup, K., Klusmann, U., Lüdtke, O., Göllner, R., & Trautwein, U. (2018). Student misbehavior and teacher well-being: Testing the mediating role of the teacher-student relationship. *Learning and Instruction, 58*, 126–136. https://doi.org/10.1016/j.learninstruc.2018.05.006

Asher, S. R., & Coie, J. D. (1990). *Peer rejection in childhood.* Cambridge University Press.

Babad, E. (1990). Measuring and changing teachers' differential behavior as perceived by students and teachers. *Journal of Educational Psychology, 82*(4), 683–690. https://doi.org/10.1037/0022-0663.82.4.683

Bagwell, C. L., & Bukowski, W. M. (2018). Friendship in childhood and adolescence: Features, effects, and processes. In W. M. Bukowski, B. Laursen, & K. H. Rubin (Eds.), *Handbook of peer interactions, relationships, and groups* (2nd ed., pp. 371–390). Guilford Press.

Bennett, N., & Blundell, D. (1983). Quantity and quality of work in rows and classroom groups. *Educational Psychology, 3*(2), 93–105. https://doi.org/10.1080/0144341830030201

Bosman, R. J., Roorda, D. L., van der Veen, I., & Koomen, H. M. (2018). Teacher-student relationship quality from kindergarten to sixth grade and students' school adjustment: A person-centered approach. *Journal of School Psychology, 68*, 177–194. https://doi.org/10.1016/j.jsp.2018.03.006

Brendgen, M., Wanner, B., & Vitaro, F. (2006). Verbal abuse by the teacher and child adjustment from Kindergarten through Grade 6. *Pediatrics, 117*, 1585–1598. https://doi.org/10.1542/peds.2005-2050

Brinkworth, M. E., McIntyre, J., Juraschek, A. D., & Gehlbach, H. (2018). Teacher-student relationships: The positives and negatives of assessing both perspectives. *Journal of Applied Developmental Psychology, 55*, 24–38. https://doi.org/10.1016/j.appdev.2017.09.002

Buhs, E. S., Ladd, G. W., & Herald, S. L. (2006). Peer exclusion and victimization: Processes that mediate the relation between peer group rejection and children's classroom engagement and achievement? *Journal of Educational Psychology, 98*(1), 1–13. https://doi.org/10.1037/0022-0663.98.1.1

Bukowski, W. M., Laursen, B., & Rubin, K. H. (Eds.). (2018). *Handbook of peer interactions, relationships, and groups*. Guilford Press.

Caprara, G. V., Barbaranelli, C., Steca, P., & Malone, P. S. (2006). Teachers' self-efficacy beliefs as determinants of job satisfaction and students' academic achievement: A study at the school level. *Journal of School Psychology, 44*(6), 473–490. https://doi.org/10.1016/j.jsp.2006.09.001

Cillessen, A. H. N. (2009). Sociometric methods. In K. H. Rubin, W. M. Bukowski, & B. Laursen (Eds.), *Handbook of peer interactions, relationships, and groups* (pp. 82–99). Guilford Press.

Cillessen, A. H. N., & Marks, P. E. L. (2011). Conceptualizing and measuring popularity. In A. H. N. Cillessen, D. Schwartz, & L. Mayeux (Eds.), *Popularity in the peer system* (pp. 25–56). Guilford Press.

Cillessen, A. H. N., & van den Berg, Y. H. M. (2012). Popularity and school adjustment. In A. M. Ryan & G. W. Ladd (Eds.), *Peer relationships and adjustment at school* (pp. 135–164). Information Age Publishing.

Coie, J. D., & Kupersmidt, J. B. (1983). A behavioral analysis of emerging social status in boys' groups. *Child Development, 54*, 1400–1416. https://doi.org/10.2307/1129803

Coie, J. D., Dodge, K. A., & Kupersmidt, J. B. (1990). Peer group behavior and social status. In S. R. Asher & J. D. Coie (Eds.), *Peer rejection in childhood* (pp. 17–59). Cambridge University Press.

Coie, J. D., Cillessen, A. H. N., Dodge, K. A., Hubbard, J. A., Schwartz, D., Lemerise, E. D., & Bateman, H. (1999). It takes two to fight: A test of relational factors and a method for assessing aggressive dyads. *Developmental Psychology, 35*, 1179–1188. https://doi.org/10.1037/0012-1649.35.5.1179

Coplan, R. J., Ooi, L. L., & Rose-Krasnor, L. (2015). Naturalistic observations of schoolyard social participation: Marker variables for socio-emotional functioning in early adolescence. *Journal of Early Adolescence, 35*, 628–650. https://doi.org/10.1037/t64290-000

Crick, N. R., & Dodge, K. A. (1994). A review and reformulation of social information-processing mechanisms in children's social adjustment. *Psychological Bulletin, 115*(1), 74. https://doi.org/10.1037/0033-2909.115.1.74

Croninger, R. G., Rice, J. K., Rathbun, A., & Nishio, M. (2007). Teacher qualifications and early learning: Effects of certification, degree, and experience on first-grade student achievement. *Economics of Education Review, 26*(3), 312–324. https://doi.org/10.1016/j.econedurev.2005.05.008

Cuadros, O., & Berger, C. (2016). The protective role of friendship quality on the wellbeing of adolescents victimized by peers. *Journal of Youth and Adolescence, 45*(9), 1877–1888. https://doi.org/10.1007/s10964-016-0504-4

Darling-Hammond, L. (2000). Teacher quality and student achievement: A review of state policy evidence. *Education Policy Analysis Archives, 8*(1), 1–44. https://doi.org/10.14507/epaa.v8n1.2000

De Laet, S., Doumen, S., Vervoort, E., Colpin, H., Van Leeuwen, K., Goossens, L., & Verschueren, K. (2014). Transactional links between teacher-child relationship quality and perceived versus sociometric popularity: A three-wave longitudinal study. *Child Development, 85*(4), 1647–1662. https://doi.org/10.1111/cdev.12216

Dijkstra, J. K., Lindenberg, S., & Veenstra, R. (2008). Beyond the class norm: Bullying behavior of popular adolescents and its relation to peer acceptance and rejection. *Journal of Abnormal Child Psychology, 36*(8), 1289–1299. https://doi.org/10.1007/s10802-008-9251-7

Dodge, K. A. (1983). Behavioral antecedents of peer social status. *Child Development, 54*, 1386–1399. https://doi.org/10.2307/1129802

Dodge, K. A. (1986). A social information processing model of social competence in children. In M. Perlmutter (Ed.), *The Minnesota symposium in child psychology* (Vol. 18, pp. 77–125). Erlbaum.

Elledge, L. C., Elledge, A. R., Newgent, R. A., & Cavell, T. A. (2016). Social risk and peer victimization in elementary school children: The protective role of teacher-student relationships. *Journal of Abnormal Child Psychology, 44*, 691–703. https://doi.org/10.1007/s10802-015-0074-z

Farmer, T. W., Lines, M. M., & Hamm, J. V. (2011). Revealing the invisible hand: The role of teachers in children's peer experiences. *Journal of Applied Developmental Psychology, 32*(5), 247–256. https://doi.org/10.1016/j.appdev.2011.04.006

Fauth, B., Decristan, J., Rieser, S., Klieme, E., & Büttner, G. (2014). Student ratings of teaching quality in primary school: Dimensions and prediction of student outcomes. *Learning and Instruction, 29*, 1–9. https://doi.org/10.1016/j.learninstruc.2013.07.001

Forbes, M. K., Fitzpatrick, S., Magson, N. R., & Rapee, R. M. (2019). Depression, anxiety, and peer victimization: Bidirectional relationships and associated outcomes transitioning from childhood to adolescence. *Journal of Youth and Adolescence, 48*(4), 692–702. https://doi.org/10.1007/s10964-018-0922-6

Furrer, C. J., Skinner, E. A., & Pitzer, J. R. (2014). The influence of teacher and peer relationships on students' classroom engagement and everyday motivation and resilience. *National Society for the Study of Education, 113*(1), 101–123.

Gerritsen, S., Plug, E., & Webbink, D. (2017). Teacher quality and student achievement: Evidence from a sample of Dutch twins. *Journal of Applied Econometrics, 32*(3), 643–660. https://doi.org/10.1002/jae.2539

Geukens, F., Maes, M., Cillessen, A. H. N., Colpin, H., van Leeuwen, K., Verschueren, K., & Goossens, L. (2021). Spotting loneliness at school: Associations between self-reports and teacher and peer nominations. *International Journal of Environmental Research and Public Health, 18*, 971. https://doi.org/10.3390/ijerph18030971

Gini, G., & Pozzoli, T. (2009). Association between bullying and psychosomatic problems: A meta-analysis. *Pediatrics, 123*(3), 1059–1065. https://doi.org/10.1542/peds.2008-1215

Gommans, R., Segers, E., Burk, W. J., & Scholte, R. H. J. (2015). The role of perceived popularity on collaborative learning: A dyadic perspective. *Journal of Educational Psychology, 107*(2), 599–608. https://doi.org/10.1037/a0037851

Gremmen, M. C., van den Berg, Y. H. M., Steglich, C., Veenstra, R., & Dijkstra, J. K. (2018). The importance of near-seated peers for elementary students' academic engagement and achievement. *Journal of Applied Developmental Psychology, 57*, 42–52. https://doi.org/10.1016/j.appdev.2018.04.004

Hamm, J. V., Farmer, T. W., Dadisman, K., Gravelle, M., & Murray, A. R. (2011). Teachers' attunement to students' peer group affiliations as a source of improved student experiences of the school social-affective context following the middle school transition. *Journal of Applied Developmental Psychology, 32*(5), 267–277. https://doi.org/10.1016/j.appdev.2010.06.003

Hamre, B. K., & Pianta, R. C. (2001). Early teacher-child relationships and the trajectory of children's school outcomes through eighth grade. *Child Development, 72*(2), 625–638. https://doi.org/10.1111/1467-8624.00301

Hamre, B. K., & Pianta, R. C. (2006). Student-teacher relationships. In G. G. Bear & K. M. Minke (Eds.), *Children's needs III: Development, prevention, and intervention* (pp. 59–71). National Association of School Psychologists.

Harris, D. N., & Sass, T. R. (2011). Teacher training, teacher quality and student achievement. *Journal of Public Economics, 95*(7–8, 798), –812. https://doi.org/10.1037/e722772011-001

Hartup, W. W. (1996). The company they keep: Friendships and their developmental significance. *Child Development, 67*, 1–13. https://doi.org/10.2307/1131681

Hartup, W. W., Laursen, B., Stewart, M. I., & Eastenson, A. (1988). Conflict and the friendship relations of young children. *Child Development, 59*, 1590–1600. https://doi.org/10.2307/1130673

Hastings, N., & Schwieso, J. (1995). Tasks and tables: The effects of seating arrangements on task engagement in primary classrooms. *Educational Research, 37*(3), 279–291. https://doi.org/10.1080/0013188950370306

Hendrickx, M. M. H. G., Mainhard, M. T., Boor-Klip, H. J., Cillessen, A. H. N., & Brekelmans, M. (2016). Social dynamics in the classroom: Teacher support and conflict and the peer ecology. *Teaching and Teacher Education, 53*, 30–40. https://doi.org/10.1016/j.tate.2015.10.004

Hendrickx, M. M. H. G., Mainhard, T., Oudman, S., Boor-Klip, H. J., & Brekelmans, M. (2017). Teacher behavior and peer liking and disliking: The teacher as a social referent for peer status. *Journal of Educational Psychology, 109*(4), 546–558. https://doi.org/10.1037/edu0000157

Hightower, A. D., Work, W. C., Cowen, E. L., Lotyczewski, B. S., Spinell, A. P., Guare, J. C., & Rohrbeck, C. A. (1986). The teacher-child rating scale: A brief objective measure of elementary children's school problem behaviors and competencies. *School Psychology Review, 15*(3), 393–409. https://doi.org/10.1080/02796015.1986.12085242

Huang, F. L., & Moon, T. R. (2009). Is experience the best teacher? A multilevel analysis of teacher characteristics and student achievement in low performing schools. *Educational Assessment, Evaluation and Accountability, 21*, 209–234. https://doi.org/10.1007/s11092-009-9074-2

Hubbard, J. A., Dodge, K. A., Cillessen, A. H. N., Coie, J. D., & Schwartz, D. (2001). The dyadic nature of social information processing in boys' reactive and proactive aggression. *Journal of Personality and Social Psychology, 80*, 268–280. https://doi.org/10.1037/0022-3514.80.2.268

Hughes, J. N., & Chen, Q. (2011). Reciprocal effects of student-teacher and student-peer relatedness: Effects on academic self efficacy. *Journal of Applied Developmental Psychology, 32*, 278–287. https://doi.org/10.1016/j.appdev.2010.03.005

Hughes, J. N., Cavell, T. A., & Willson, V. (2001). Further support for the developmental significance of the quality of the teacher-student relationship. *Journal of School Psychology, 39*(4), 289–301. https://doi.org/10.1016/S0022-4405(01)00074-7

Hughes, J. N., Im, M. H., & Wehrly, S. E. (2014). Effect of peer nominations of teacher-student support at individual and classroom levels on social and academic outcomes. *Journal of School Psychology, 52*, 309–322. https://doi.org/10.1016/j.jsp.2013.12.004

Huitsing, G., Veenstra, R., Sainio, M., & Salmivalli, C. (2012). "It must be me" or "It could be them?": The impact of the social network position of bullies and victims on victims' adjustment. *Social Networks, 34*(4), 379–386. https://doi.org/10.1016/j.socnet.2010.07.002

Huitsing, G., Lodder, G. M., Oldenburg, B., Schacter, H. L., Salmivalli, C., Juvonen, J., & Veenstra, R. (2019). The healthy context paradox: Victims' adjustment during an anti-bullying intervention. *Journal of Child and Family Studies, 28*(9), 2499–2509. https://doi.org/10.1007/s10826-018-1194-1

Jerome, E. M., Hamre, B. K., & Pianta, R. C. (2009). Teacher-child relationships from Kindergarten to sixth grade: Early childhood predictors of teacher-perceived conflict and closeness. *Social Development, 18*, 915–945. https://doi.org/10.1111/j.1467-9507.2008.00508.x

Kenny, D. A., Kashy, D. A., & Cook, W. L. (2006). *Dyadic data analysis*. Guilford Press.

Koomen, H. M., & Jellesma, F. C. (2015). Can closeness, conflict, and dependency be used to characterize students' perceptions of the affective relationship with their teacher? Testing a new child measure in middle childhood. *British Journal of Educational Psychology, 85*(4), 479–497. https://doi.org/10.1111/bjep.12094

Koomen, H. M., Verschueren, K., van Schooten, E., Jak, S., & Pianta, R. C. (2012). Validating the student-teacher relationship scale: Testing factor structure and measurement invariance across child gender and age in a Dutch sample. *Journal of School Psychology, 50*(2), 215–234. https://doi.org/10.1016/j.jsp.2011.09.001

Ladd, G. W. (2006). Peer rejection, aggressive or withdrawn behavior, and psychological maladjustment from ages 5 to 12: An examination of four predictive models. *Child Development, 77*, 822–846. https://doi.org/10.1111/j.1467-8624.2006.00905.x

Lansford, J. E., Malone, P. S., Dodge, K. A., Pettit, G. S., & Bates, J. E. (2010). Developmental cascades of peer rejection, social information processing biases, and aggression during middle childhood. *Development and Psychopathology, 22*(3), 593–602. https://doi.org/10.1017/s0954579410000301

Lansu, T. A. M., van Noorden, T. H. J., & Deutz, M. H. F. (2017). How children's victimization relates to distorted versus sensitive social cognition: Perception, mood, and need fulfillment in response to Cyberball inclusion and exclusion. *Journal of Experimental Child Psychology, 154*, 131–145. https://doi.org/10.1016/j.jecp.2016.10.012

Lei, H., Cui, Y., & Chiu, M. M. (2016). Affective teacher-student relationships and students' externalizing behavior problems: A meta-analysis. *Frontiers in Psychology, 7*, 1–12. https://doi.org/10.3389/fpsyg.2016.01311

Lucas-Molina, B., Williamson, A. A., Pulido, R., & Pérez-Albéniz, A. (2015). Effects of teacher-student relationships on peer harassment: A multilevel study. *Psychology in the Schools, 52*, 298–315. https://doi.org/10.1002/pits.21822

Marx, A., Fuhrer, U., & Hartig, T. (1999). Effects of classroom seating arrangements on children's question-asking. *Learning Environments Research, 2*(3), 249–263. https://doi.org/10.1023/a:1009901922191

McAuliffe, M. D., Hubbard, J. A., & Romano, L. J. (2009). The role of teacher cognition and behavior in children's peer relations. *Journal of Abnormal Child Psychology, 37*(5), 665–677. https://doi.org/10.1007/s10802-009-9305-5

McDonald, K. L., & Asher, S. R. (2018). Peer acceptance, peer rejection, and popularity: Social-cognitive and behavioral perspectives. In W. M. Bukowski, B. Laursen, & K. H. Rubin (Eds.), *Handbook of peer interactions, relationships, and groups* (2nd ed., pp. 429–446). Guilford Press.

McKown, C., & Weinstein, R. S. (2008). Teacher expectations, classroom context, and the achievement gap. *Journal of School Psychology, 46*(3), 235–261. https://doi.org/10.1016/j.jsp.2007.05.001

Moore, B., & Woodcock, S. (2017). Resilience, bullying, and mental health: Factors associated with improved outcomes. *Psychology in the Schools, 54*(7), 689–702. https://doi.org/10.1002/pits.22028

Newcomb, A. F., Bukowski, W. M., & Pattee, L. (1993). Children's peer relations: A meta-analytic review of popular, rejected, neglected, controversial, and average sociometric status. *Psychological Bulletin, 113*(1), 99. https://doi.org/10.1037/0033-2909.113.1.99

Norwalk, K. E., Hamm, J. V., Farmer, T. W., & Barnes, K. L. (2016). Improving the school context of early adolescence through teacher attunement to victimization: Effects on school belonging. *Journal of Early Adolescence, 36*(7), 989–1009. https://doi.org/10.1177/0272431615590230

Oldenburg, B., van Duijn, M., Sentse, M., Huitsing, G., van der Ploeg, R., Salmivalli, C., & Veenstra, R. (2015). Teacher characteristics and peer victimization in elementary schools: A classroom-level perspective. *Journal of Abnormal Child Psychology, 43*(1), 33–44. https://doi.org/10.1007/s10802-013-9847-4

Oldenburg, B., Bosman, R., & Veenstra, R. (2016). Are elementary school teachers prepared to tackle bullying? A pilot study. *School Psychology International, 37*(1), 64–72. https://doi.org/10.1177/0143034315623324

Olweus, D. (1994). Bullying at school: Basic facts and effects of a school based intervention program. *Journal of Child Psychology and Psychiatry, 35*(7), 1171–1190. https://doi.org/10.1111/j.1469-7610.1994.tb01229.x

Parkhurst, J. T., & Hopmeyer, A. (1998). Sociometric popularity and peer-perceived popularity: Two distinct dimensions of peer status. *Journal of Early Adolescence, 18*(2), 125–144. https://doi.org/10.1177/0272431698018002001

Perren, S., Ettekal, I., & Ladd, G. W. (2013). The impact of peer victimization on later maladjustment: Mediating and moderating effects of hostile and self-blaming attributions. *Journal of Child Psychology and Psychiatry, 54*, 46–55. https://doi.org/10.1111/j.1469-7610.2012.02618.x

Peters, E., Cillessen, A. H. N., Riksen-Walraven, J. M., & Haselager, G. J. T. (2010). Best friends' preference and popularity: Associations with aggression and prosocial behavior. *International Journal of Behavioral Development, 34*(5), 398–405. https://doi.org/10.1177/0165025409343709

Peters, E., Riksen-Walraven, J. M., Cillessen, A. H. N., & de Weerth, C. (2011). Peer rejection and HPA activity in middle childhood: Friendship makes a difference. *Child Development, 82*(6), 1906–1920. https://doi.org/10.1111/j.1467-8624.2011.01647.x

Pianta, R. C. (2001). *Student-teacher relationship scale: Professional manual*. Psychological Assessment Resources.

Pianta, R. C., & Stuhlman, M. W. (2004). Teacher-child relationships and children's success in the first years of school. *School Psychology Review, 33*(3), 444–458. https://doi.org/10.1080/02796015.2004.12086261

Pouwels, J. L., van Noorden, T. H. J., Lansu, T. A. M., & Cillessen, A. H. N. (2018). The participant roles of bullying in different grades: Prevalence and social status profiles. *Social Development, 27*(4), 732–747. https://doi.org/10.1111/sode.12294

Prinstein, M. J., Rancourt, D., Adelman, C. B., Ahlich, E., Smith, J., & Guerry, J. D. (2018). Friendship in childhood and adolescence: Features, effects, and processes. In W. M. Bukowski, B. Laursen, & K. H. Rubin (Eds.), *Handbook of peer interactions, relationships, and groups* (2nd ed., pp. 617–636). Guilford Press.

Reijntjes, A., Kamphuis, J. H., Prinzie, P., & Telch, M. J. (2010). Peer victimization and internalizing problems in children: A meta-analysis of longitudinal studies. *Child Abuse & Neglect, 34*(4), 244–252. https://doi.org/10.1016/j.chiabu.2009.07.009

Rodkin, P. C., Farmer, T. W., Pearl, R., & Van Acker, R. (2000). Heterogeneity of popular boys: Antisocial and prosocial configurations. *Developmental Psychology, 36*(1), 14–24. https://doi.org/10.1037/0012-1649.36.1.14

Roorda, D. L., Koomen, H. M., Spilt, J. L., & Oort, F. J. (2011). The influence of affective teacher-student relationships on students' school engagement and achievement: A meta-analytic approach. *Review of Educational Research, 81*(4), 493–529. https://doi.org/10.3102/0034654311421793

Roorda, D. L., Jak, S., Zee, M., Oort, F. J., & Koomen, H. M. (2017). Affective teacher-student relationships and students' engagement and achievement: A meta-analytic update and test of the mediating role of engagement. *School Psychology Review, 46*(3), 239–261. https://doi.org/10.17105/spr-2017-0035.v46-3

Rosenfield, P., Lambert, N. M., & Black, A. (1985). Desk arrangement effects on pupil classroom behavior. *Journal of Educational Psychology, 77*(1), 101. https://doi.org/10.1037/0022-0663.77.1.101

Rosenthal, R., & Jacobson, L. (1968). Pygmalion in the classroom. *The Urban Review, 3*(1), 16–20. https://doi.org/10.1007/bf02322211

Rubie-Davies, C. M. (2007). Classroom interactions: Exploring the practices of high-and low-expectation teachers. *British Journal of Educational Psychology, 77*(2), 289–306. https://doi.org/10.1348/000709906x101601

Rubie-Davies, C., Hattie, J., & Hamilton, R. (2006). Expecting the best for students: Teacher expectations and academic outcomes. *British Journal of Educational Psychology, 76*(3), 429–444. https://doi.org/10.1348/000709905x53589

Rubie-Davies, C., Meissel, K., Alansari, M., Watson, P., Flint, A., & McDonald, L. (2020). Achievement and beliefs outcomes of students with high and low expectation teachers. *Social Psychology of Education, 23*(5), 1173–1201. https://doi.org/10.1007/s11218-020-09574-y

Sabol, T. J., & Pianta, R. C. (2012). Recent trends in research on teacher-child relationships. *Attachment & Human Development, 14*(3), 213–231. https://doi.org/10.1080/14616734.2012.672262

Sainio, M., Veenstra, R., Huitsing, G., & Salmivalli, C. (2011). Victims and their defenders: A dyadic approach. *International Journal of Behavioral Development, 35*(2), 144–151. https://doi.org/10.1177/0165025410378068

Salmivalli, C., Lagerspetz, K., Björkqvist, K., Österman, K., & Kaukiainen, A. (1996). Bullying as a group process: Participant roles and their relations to social status within the group. *Aggressive Behavior, 22*(1), 1–15. https://doi.org/10.1002/(sici)1098-2337

Sandstrom, M. J., & Cillessen, A. H. N. (2003). Sociometric status and children's peer experiences: Use of the daily diary method. *Merrill-Palmer Quarterly, 49*, 427–452.

Sandstrom, M. J., Deutz, M. H., Lansu, T. A., van Noorden, T. H., Karremans, J. C., & Cillessen, A. H. N. (2017). Unanimous versus partial rejection: How the number of excluders influences the impact of ostracism in children. *Aggressive Behavior, 43*(2), 190–203. https://doi.org/10.1002/ab.21674

Schwartz, D., Gorman, A. H., Nakamoto, J., & McKay, T. (2006). Popularity, social acceptance, and aggression in adolescent peer groups: Links with academic performance and school attendance. *Developmental Psychology, 42*(6), 1116–1127. https://doi.org/10.1037/0012-1649.42.6.1116

Sentse, M., Kretschmer, T., & Salmivalli, C. (2015). The longitudinal interplay between bullying, victimization, and social status: Age-related and gender differences. *Social Development, 24*(3), 659–677. https://doi.org/10.1111/sode.12115

Skaalvik, E. M., & Skaalvik, S. (2007). Dimensions of teacher self-efficacy and relations with strain factors, perceived collective teacher efficacy, and teacher burnout. *Journal of Educational Psychology, 99*(3), 611. https://doi.org/10.1037/0022-0663.99.3.611

Spilt, J. L., Koomen, H. M., & Thijs, J. T. (2011). Teacher wellbeing: The importance of teacher-student relationships. *Educational Psychology Review, 23*(4), 457–477. https://doi.org/10.1007/s10648-011-9170-y

Stronge, J. H., Ward, T. J., Tucker, P. D., & Hindman, J. L. (2007). What is the relationship between teacher quality and student achievement? An exploratory study. *Journal of Personnel Evaluation in Education, 20*(3), 165–184. https://doi.org/10.1007/s11092-008-9053-z

Troop-Gordon, W. (2015). The role of the classroom teacher in the lives of children victimized by peers. *Child Development Perspectives, 9*(1), 55–60. https://doi.org/10.1111/cdep.12106

Troop-Gordon, W., & Ladd, G. W. (2015). Teachers' victimization-related beliefs and strategies: Associations with students' aggressive behavior and peer victimization. *Journal of Abnormal Child Psychology, 43*(1), 45–60. https://doi.org/10.1007/s10802-013-9840-y

van den Berg, Y. H. M., & Cillessen, A. H. N. (2015). Peer status and classroom seating arrangements: A social relations analysis. *Journal of Experimental Child Psychology, 130*, 19–34. https://doi.org/10.1016/j.jecp.2014.09.007

van den Berg, Y. H. M., & Stoltz, S. E. M. J. (2018). Enhancing social inclusion of children with externalizing problems through classroom seating arrangements: A randomized controlled trial. *Journal of Emotional and Behavioral Disorders, 26*(1), 31–41. https://doi.org/10.1177/1063426617740561

van den Berg, Y. H. M., Segers, E., & Cillessen, A. H. N. (2012). Changing peer perceptions and victimization through classroom arrangements: A field experiment. *Journal of Abnormal Child Psychology, 40*(3), 403–412. https://doi.org/10.1007/s10802-011-9567-6

van der Ploeg, R., Kretschmer, T., Salmivalli, C., & Veenstra, R. (2017). Defending victims: What does it take to intervene in bullying and how is it rewarded by peers? *Journal of School Psychology, 65*, 1–10. https://doi.org/10.1016/j.jsp.2017.06.002

van der Ploeg, R., Stoltz, S., van den Berg, Y. H. M., Cillessen, A. H. N., & Orobio de Castro, B. (2021). *To disclose or not? Children's tendency to disclose peer victimization in elementary school*. Manuscript under review.

Veenstra, R., Lindenberg, S., Huitsing, G., Sainio, M., & Salmivalli, C. (2014). The role of teachers in bullying: The relation between antibullying attitudes, efficacy, and efforts to reduce bullying. *Journal of Educational Psychology, 106*(4), 1135. https://doi.org/10.1037/a0036110

Verschueren, K., & Koomen, H. M. (2012). Teacher-child relationships from an attachment perspective. *Attachment & Human Development, 14*(3), 205–211. https://doi.org/10.1080/14616734.2012.672260

Visconti, K. J., Kochenderfer-Ladd, B., & Clifford, C. A. (2013). Children's attributions for peer victimization: A social comparison approach. *Journal of Applied Developmental Psychology, 34*(6), 277–287. https://doi.org/10.1016/j.appdev.2013.06.002

Wang, W., Vaillancourt, T., Brittain, H. L., McDougall, P., Krygsman, A., Smith, D., … Hymel, S. (2014). School climate, peer victimization, and academic achievement: Results from a multi-informant study. *School Psychology Quarterly, 29*(3), 360. https://doi.org/10.1037/e574242014-057

Wang, S., Rubie-Davies, C. M., & Meissel, K. (2020). The stability and trajectories of teacher expectations: Student achievement level as a moderator. *Learning and Individual Differences, 78*, 1–10. https://doi.org/10.1016/j.lindif.2019.101819

Wannarka, R., & Ruhl, K. (2008). Seating arrangements that promote positive academic and behavioral outcomes: A review of empirical research. *Support for Learning, 23*(2), 89–93. https://doi.org/10.1111/j.1467-9604.2008.00375.x

Wubbels, T., Brekelmans, M., Den Brok, P., & van Tartwijk, J. (2006). An interpersonal perspective on classroom management in secondary classrooms in the Netherlands. In C. Evertson & C. Weinstein (Eds.), *Handbook of classroom management: Research, practice and contemporary issues* (pp. 1161–1191). Erlbaum.

Wubbels, T., Brekelmans, M., Den Brok, P., Wijsman, L., Mainhard, T., & van Tartwijk, J. (2014). Teacher-student relationships and classroom management. In E. T. Emmer & E. J. Sabornie (Eds.), *Handbook of classroom management* (2nd ed., pp. 363–386). Routledge.

Yang, C., Sharkey, J. D., Reed, L. A., Chen, C., & Dowdy, E. (2018). Bullying victimization and student engagement in elementary, middle, and high schools: Moderating role of school climate. *School Psychology Quarterly, 33*(1), 54. https://doi.org/10.1037/spq0000250

Yeager, D. S., Johnson, R., Spitzer, B. J., Trzesniewski, K. H., Powers, J., & Dweck, C. S. (2014). The far-reaching effects of believing people can change: Implicit theories of personality shape stress, health, and achievement during adolescence. *Journal of Personality and Social Psychology, 106*, 867–884. https://doi.org/10.1037/a0036335

Yoon, J. S., & Kerber, K. (2003). Bullying: Elementary teachers' attitudes and intervention strategies. *Research in Education, 69*(1), 27–35. https://doi.org/10.7227/rie.69.3

Zucchetti, G., Candela, F., Sacconi, B., & Rabaglietti, E. (2015). Friendship quality and school achievement: A longitudinal analysis during primary school. *Journal of Applied School Psychology, 31*, 297–314. https://doi.org/10.1080/15377903.2015.1084963

How Peers and Teachers Shape Elementary School Children's Academic and Socioemotional Development

Frank Vitaro, Hilde Colpin, Jacintha Tieskens, and Pol A. C. van Lier

Introduction

Throughout the elementary school years, children tend to spend more time with their peers and teachers than with their parents. Based on bio-ecological models of development (e.g., Bronfenbrenner & Morris, 2006), these outside-of-the-family experiences ought to contribute, for better or worse, to their socioemotional and academic development. There is, indeed, empirical evidence showing that relations with peers and teachers predict different types of student outcomes. However, limitations in study design often interfere with the proper identification of such influences because not all methodologies (e.g., cross-sectional designs) effectively maintain internal validity and establish directionality of influence. In addition, results across studies may differ because experiences with peers and with teachers may play different roles (e.g., risk/beneficial factors, moderators, or mediators) depending on the specific type of experience at play (e.g., friendship participation vs friends' characteristics) and the different outcomes under scrutiny (e.g., internalizing vs externalizing problems) (Vitaro, Boivin, & Bukowski, 2009a). Results may also differ depending on participants' characteristics (e.g., age, sex, and socio-behavioral profile). Therefore, to best depict a clear picture of the roles that peers and teachers play in children's lives, we adopted a broad scope that includes

F. Vitaro (✉)
School of Psychoeducation, University of Montreal, Montreal, Canada
e-mail: frank.vitaro@umontreal.ca

H. Colpin
School Psychology and Development in Context Research Unit, KU Leuven University, Leuven, Belgium
e-mail: hilde.colpin@kuleuven.be

J. Tieskens · P. A. C. van Lier
Clinical Developmental Psychology, Vrije Universiteit Amsterdam, Amsterdam, Netherlands
e-mail: j.m.tieskens@vu.nl; pac.van.lier@vu.nl

P. A. C. van Lier, K. Deater-Deckard (eds.), *Biosocial Interplay During Elementary School*, https://doi.org/10.1007/978-3-031-07109-6_3

different types of social experiences and different outcomes while limiting the developmental period under scrutiny (i.e., to childhood/elementary school period). More specifically, we focused on two types of peer processes: those at the dyadic level (i.e., friendship participation, friends' characteristics, friendship quality) and those at the group level (i.e., peer rejection/acceptance, peer victimization, peer group norms). Most developmental theorists emphasize the importance of both types of peer experiences as antecedents of multiple forms of behavioral adjustment, including socioemotional development and school engagement (Kindermann, 2016; Sullivan, 1953). We also covered two parallel types of teacher experiences: those at the group level (teacher vis-à-vis the class group, with a focus on management style) and those at the dyadic level (teacher vis-à-vis one student, with a focus on the affective quality of the teacher-child relationship). Together, these two types of teacher experiences cover most of the daily teacher-child interactions in the classroom. In line with our broad perspective, each type of peer and teacher experience was examined (a) in reference to socioemotional outcomes, such as internalizing behaviors (i.e., anxiety, withdrawal, feelings of loneliness, depressive symptoms, well-being), externalizing behaviors (i.e., aggression, opposition, hyperactivity-inattention, antisociality), and prosociality, and (b) in reference to academic outcomes such as school engagement and academic performance. We considered each type of peer and teacher experience as a predictor with a main, mediating, or interactive "effect" in regard to child's functioning in multiple domains. As much as possible, we documented the intra- or interpersonal mechanisms that could account for these different roles. However, we did not document possible reciprocal associations between domains of child functioning and changes in experiences with peers and with teachers, nor did we document factors that could mitigate or exacerbate the contribution of peer and teacher experiences (except for children's sex and age, when available). Finally, we conclude our review with a brief examination of the possible interplay between peers and teachers and a list of questions/issues for the future.

In line with our first point above, we focused on studies with strong internal validity. More specifically, we selected longitudinal studies that included strong control of third variables, studies using an experimental design, and studies using a genetically informed design. In and of itself, well-controlled longitudinal and genetically informed studies cannot provide direct and conclusive proof of causation to the extent that an experimental design can. However, we deemed it necessary to include these two types of non-experimental designs for the following reasons. First, studies using an experimental design are scarce and often limited in the type of experiences that could be manipulated (for ethical and practical reasons). Second, well-controlled longitudinal studies using, for example, a cross-lagged design and genetically informed studies have substantial virtues of their own. More specifically, well-controlled longitudinal studies using a cross-lagged design allow for the control of concurrent and stability links of both the predictors and the outcomes, in addition to the control of third variables. Longitudinal studies also allow researchers to examine the accumulated or differential "effect" of social experiences over several months or years. In contrast, experimental studies are often limited with respect to the duration of the exposure to the manipulated social experience. In a similar vein, genetically informed studies (i.e., twin studies) identify, but also account for, genetic factors inherent to the child that might be confounded with social experiences through a process known as a gene-environment correlation (rGE) or may interact with specific social experiences through a process known as a gene-environment interaction (GxE) (see Brendgen et al., 2012).

Peer Experiences at the Dyad Level

As suggested by Hartup (1996, 2005), three aspects of friendships were examined in order to more fully understand how and when friendships make a positive or a negative contribution to children's psychosocial and academic development: friendship participation, friends' characteristics, and features of friendships.

Friendship Participation Friendship participation has been typically defined as having at least one mutual friendship with another child (Bukowski & Hoza, 1989; Parker & Asher, 1993). This dichotomous view of friendship participation was derived from the finding that the number of friends is not linearly related to individual adjustment: Whereas children with no mutual friends report more internalizing symptoms than those with one or more mutual friends, those with more than one mutual friend do not differ from those with one mutual friend in that respect (Parker & Seal, 1996).

Several studies have shown that the presence/absence of at least one mutual friend during elementary school is significantly related to later adjustment. However, in accordance with our concern for internal validity, few studies have included the appropriate controls to ensure that the role of friendship participation was not spurious. A first set of variables to control is children's initial social-cognitive, behavioral, or emotional characteristics or states. Children who have friends tend to differ from children without friends on these characteristics (Newcomb & Bagwell, 1995). A second set of variables to control are correlated social experiences: compared to children lacking friends, friended children may also be exposed to a variety of other experiences due to their personal characteristics (e.g., social acceptance at the group level).

Only a few studies have included the above controls in examining the main contribution of friendship participation to child development. In one study, fifth graders who had a stable best friend viewed themselves more positively and reported fewer depressive feelings in early adulthood than those who were friendless (Bagwell et al., 1998). Of note, the benefits of having a friend were found to be specific to emotional well-being; peer acceptance, but not friendship status, predicted school performance. Another study found that the lack of a close friend from grade 1 through grade 3 predicted greater feelings of loneliness and anxious-depressed behaviors 1 year later, above and beyond initial levels of these internalized problems, peer rejection, and peer victimization (Ladd & Troop-Gordon, 2003). Using a cross-lagged design spanning from early childhood to early adolescence, Pedersen et al. (2007) also found that friendship participation predicted lower levels of loneliness and depressed feelings, even after accounting for personal and social factors such as disruptiveness, anxiety, and peer rejection. Friendship participation also partially mediated the links between these personal and social factors and depressive feelings and loneliness. However, friendship participation did not predict externalized (i.e., delinquent) behaviors. The few studies that included sex as a moderator found that the association between friendship participation and emotional well-being is especially strong for girls in the fifth and eighth grades (e.g., Oldenburg & Kerns, 1997), which could reflect a greater orientation toward, and dependence upon, social relationships in females than in males (Archer & Lloyd, 2002). However, this last finding should be qualified further by age, as these sex differences were not found among young school-age children (e.g., Ladd & Troop-Gordon, 2003).

Having at least one close friend is not only beneficial to a child's well-being; it may also protect against the negative emotional consequences of aversive social experiences (i.e., it may operate as a moderator). For example, unlike classmates who are rejected by the peer group and have no friends, school children who are rejected but who have at least one mutual friend do not report greater levels of loneliness and depression when compared to their more accepted peers (Parker & Asher, 1993). This protective effect of friendship participation in reference to peer rejection is already operative during the early school years (Laursen et al., 2007). Having at least one close friend also protects children against the continuation of peer victimization and its negative consequences. For example, a study of fourth and fifth graders who were victimized revealed that having a reciprocal best friend significantly reduced the likelihood of being re-victimized over a 1-year period (Hodges et al., 1999). It also reduced the likelihood of developing internalizing problems compared to those without a best friend. Having a best friend present during a difficult event also appears to reduce stress, as measured by the hormone cortisol (Adams et al., 2011). Similarly, possessing high-quality friendships offers protection against anxiety disorders for victims of serious abuse, such as childhood sexual abuse (Adams & Bukowski, 2007). Finally, there is evidence from genetically informed studies that friendship participation can reduce the expression of a genetic vulnerability for depression in girls (i.e., it operates as a moderator relative to genetic risk). For boys, friendship participation (and genetic vulnerability) appears to be directly related (i.e., via main effects) to depressive symptoms (Brendgen et al., 2013a).

Many processes could underlie the longitudinal association between participation in friendship and later emotional well-being. For instance, friendship participation may provide children with a "secure base" necessary to create comfort and willingness to explore new environments and get involved in new social situations (Birch & Ladd, 1996). The absence of such a secure base, in turn, can generate anxiety and lack of self-confidence. It may also negatively affect children's self-perceptions, which are central to emotional well-being and self-confidence (Ladd & Troop-Gordon, 2003). Friended children may also be less at risk for feelings of distress than solitary individuals because friends provide important social provisions such as companionship, emotional support, intimacy, and self-validation (Boivin et al., 2001). In sum, friendship participation seems to play a beneficial or protective role regarding internalizing behaviors and well-being through a number of possible processes, but does it play similar roles with respect to school achievement and externalizing behaviors?

Several short-term longitudinal studies reveal that school children who have a reciprocated friendship by the end of elementary school manifest increased levels of school liking and prosocial behavior (Erath et al., 2008; Wentzel et al., 2004). Friendships may provide motivational and instrumental support to engage and succeed in school-related activities (Erath et al., 2008) and for behaving prosocially (Wentzel et al., 2004). However, as noted earlier, friendship participation per se does not seem to affect academic performance directly (Bagwell et al., 1998). The evidence of a beneficial effect of friendship participation with respect to externalizing behaviors is also mixed. To illustrate, Laursen and colleagues found that first grade children with at least one mutual friend experienced a decrease in both internalizing *and* externalizing problems over a 1-year period compared to children without a friend (Laursen et al., 2007). In contrast, Vitaro and collaborators found that disruptive boys with at least one mutual friend remained disruptive over a 1-year period compared to disruptive boys with no friends (Vitaro et al., 1997). One possible way to reconcile these contradictory findings is to consider the characteristics of the friends involved in these friendships.

Friends' Characteristics Friendship participation can be beneficial or protective if the friends are well adjusted. For example, evidence from longitudinal data show that affiliation with prosocial friends is related to an increase in children's prosocial behavior and can also reduce children's risk of having problematic relations with other peers (Barry & Wentzel, 2006; Eivers et al., 2012; Wentzel et al., 2004). However, friendships can be detrimental if the friends are deviant. There is indeed strong evidence to suggest that friends' externalizing problems (e.g., antisociality, aggression) predict the maintenance and the escalation of similar externalized behavior problems during the elementary school years. To illustrate, using a cross-lagged design, Buil and her collaborators showed that friends' antisocial behavior from age 8 through age 13 positively predicted overt (i.e., aggression) and covert (i.e., vandalism and theft) antisocial behavior during adolescence, net of the children's personal dispositions (i.e., temperament and antisocial behavior) and of their social status at the group level (i.e., poor social preference) (Buil et al., 2017). Results from other studies suggest that the negative consequences of exposure to deviant friends may be apparent as early as young childhood. For example, Snyder and his collaborators demonstrated that the affiliation with deviant peers in kindergarten predicts growth in overt conduct problems (e.g., aggressiveness) as well as covert conduct problems (e.g., lying and stealing) during the following 2 years (Snyder et al., 2005). It is important to note, however, that the "influence" of friends' externalized problems at this young age may be very specific. To illustrate, Lamarche et al. (2007) found that friends' reactive aggression specifically predicted increases in participants' reactive aggression, whereas friends' proactive aggression specifically predicted increases in participants' proactive aggression from kindergarten to grade 1, but not the other way around (reactive aggression refers to aggressive outburst following threats or insults, whereas proactive aggression corresponds to aggressive acts manifested without provocation for personal gain).

The available evidence about the developmental role of friends' characteristics rests mainly on correlational, but albeit longitudinal, studies that did not control for possible third variables such as family and genetic factors, forcing the use of cautionary brackets when using causal terms such as friends' "influence." Indeed, there is evidence showing that socio-family and genetic factors may affect social experiences, not only behavior, and thus operate as third unaccounted common factors; for genetic factors, the processes at play are known as gene-environment correlations, or rGE (see Brendgen et al., 2012). Fortunately, the application of methodological strategies such as the monozygotic (MZ) twin-difference method allows one to control for likely genetic and family-wide influences. Since MZ twins from the same pair not only share 100% of their genes but also share the same socio-familial environment when raised together, any behavioral differences between them are attributed to nonshared environmental influences (see Vitaro, Brendgen, & Arseneault, 2009b, for a full description of the method). When adapted to investigations regarding the influence of friends' characteristics (or other social experiences) on child behaviors, this method bolsters confidence that identified associations are not spurious due to genetic and environmental overlap. As an example, Vitaro et al. (2011) found that MZ-twin within-pair differences in friends' physical aggression at age 6 years predicted an increase in MZ-twin within-pair differences in physical aggression from age 6 to 7 years (Vitaro et al., 2011). Using a similar design and similar measures, the same group of researchers did not find however a link between MZ-twin within-pair differences in friends' physical aggression at age 10 and MZ-twin within-pair differences in twins' physical aggression at age 13, possibly because of the more important role played by rGE during late childhood compared to early childhood (Vitaro et al., 2016).

Friends' characteristics may also moderate genetic influences on the expression of children's problem behaviors during the elementary school years. For example, several genetically informed studies demonstrate that an individual's genetic liability toward externalized problems may be expressed more or less strongly as a function of affiliating with aggressive peers (i.e., an example of a gene-environment interaction, or GxE). For example, 6-year-old children enrolled in kindergarten displayed higher levels of aggression if they were at high genetic risk for such behavior *and* were also exposed to highly aggressive friends (van Lier et al., 2007). A follow-up study in first grade revealed that this same gene-by-environment interaction held for physical aggression, but not for social aggression (Brendgen et al. 2008a).

Researchers have proposed several explanatory mechanisms at the interpersonal level to account for the main or moderating effects of friends' "negative" characteristics on children's problematic behaviors. A first process, labeled *deviancy training*, has received substantial empirical support. Specifically, deviant friends tend to reinforce (through laughter or positive nonverbal feedback) rule-breaking talk or deviant acts and tend to ignore or punish normative behaviors (Dishion et al., 1996). Deviancy training may occur among kindergarten children; in one study, engaging in deviant talk and imitative play of deviant behaviors with same-gender peers predicted an increase in overt and covert conduct problems on the playground, at school, and at home (Snyder et al., 2005). A second process that may facilitate deviancy training in the context of friendships is *pressure to conform to norm-breaking (or norm-obedience) behaviors*. For example, when compared to non-aggressive boys and their friends, 10-year-old aggressive boys and their friends tend to provide more enticement for rule violations in situations that provided opportunities for rule-breaking behavior (Bagwell & Coie, 2004). *Demonstration-imitation through observational learning* of rule-breaking or aggressive behaviors is a third process that may also explain how friends support the acquisition, maintenance, or escalation of aggressive-antisocial behavior (Berndt, 1999; Hartup & Stevens, 1997). As shown by Thompson et al. (2019), all the above processes can simultaneously and independently play a role. More specifically, these authors found reciprocal relations between peers' problem behavior, peer pressure for fighting, and peers' support for fighting and changes in self-reported aggressive behavior from age 10 through age 16. The positive counterparts of these processes (i.e., conformity training, motivation to conform to prosocial norms through feelings of retribution, or demonstration-imitation) can also explain how friends' prosocial behavior by late childhood predicts change in individuals' prosocial behavior (Wentzel et al., 2004).

Antisocial children tend to be bossier with their friends and are often more frequently involved in coercive and conflictual exchanges than conventional children (Deptula & Cohen, 2004). These conflictual-negative interactions could set in motion a *coercive interactional process* (Boivin & Vitaro, 1995) whereby coercing or threatening one's friend for some personal benefit, if successful, can increase the likelihood of similar coercive behaviors in the future through negative reinforcement. Consistent with this notion, coercion from a best friend accounted for the link between friends' aggression and an increase in participants' aggression in young boys (Vitaro et al., 2011). Importantly, it is possible that different processes underlie friends' "influence" depending on the type of outcome at stake. For example, there is evidence showing that coercion is involved with aggression-type outcomes, whereas deviancy training and modeling foster externalizing problems of the covert type such as stealing and cheating (Snyder et al., 2007). Coercion from an aggressive friend can also affect children's depressive mood (Vitaro et al., 2011).

However, friends do not need to be aggressive or deviant to affect children's mood or internal feelings. There is accumulating evidence that friends' internalizing problems can foster children's internalizing problems such as anxiety and depressive feelings, especially among girls (Prinstein, 2007; Stevens & Prinstein, 2005). Negative affect can spread between friends through co-rumination, a form of disclosure that involves rehashing and excessively discussing problems, mutual encouragement of problem talk, and dwelling on negative affect (Rose, 2002). The vast majority of studies examining this topic included adolescents. There is one study, however, that found that co-rumination among friends in late childhood predicts increased internalizing problems, specifically anxiety, while accounting for possible rGE through the use of the MZ-twin difference method (Dirghangi et al., 2015).

By increasing children's externalized or internalized problems, friends' externalized or internalized behavior can ultimately, albeit indirectly, negatively influence children's academic success or school motivation. Such indirect effects of friends' characteristics regarding school-related outcomes lack empirical evidence at this moment. However, there is evidence that friends' academic behavior can directly influence children's academic achievement. For example, Nelson and DeBacker (2008) found that having a best friend who values academics positively contributes to children's adaptive achievement motivation. Similarly, Kindermann and Skinner (2012) found that the engagement levels of friends predicted changes in children's classroom engagement, independent of levels of engagement of the rest of classmates (which also made a significant contribution). Conversely, students who associate with friends who reject school are more likely to perform poorly academically (Véronneau et al., 2008) and more likely to drop out of school (Fergusson et al., 1999). However, when controlling for both current peer acceptance at the group level and autoregressive effects through the use of a cross-lagged design from grade 2 to grade 7, Véronneau and her colleagues did not find a predictive link between friends' academic achievement and changes in participants' academic achievement. Only peer rejection at the group level predicted decreases in academic achievement from grades 3 through 5 (Véronneau et al., 2010). In sum, the role friends play with respect to children's academic achievement by virtue of their personal characteristics is yet unsettled. This might be because the role of friends' characteristics may depend on the quality of the friendship.

Friendship Quality Unfortunately, studies examining friendship quality as a moderator of the link between friends' characteristics and changes in children's behavior or academic performance are scarce; they are also inconclusive. To illustrate, in one study, high friendship quality (i.e., caring, sharing, helping, laughing) mitigated the link between friends' aggression and changes in young children's aggression (Salvas et al., 2011). Yet, in other studies involving young adolescents, high friendship quality exacerbated deviant friends' "influence" on young adolescents' externalizing behaviors (Piehler & Dishion, 2007). Finally, in a third study, it was low- (i.e., conflict, betrayal, coercion), not high-, quality friendship that exacerbated the risk associated with exposure to deviant or depressed friends (Poulin et al., 1999; Prinstein, 2007).

The evidence regarding the main effect of friendship quality in regard to children's socioemotional development is more consistent. To illustrate, a child's involvement in an intimate, trustworthy, caring, and supportive friendship positively predicts his/her well-being and prosocial behavior (e.g., Betts & Rotenberg, 2007; Ladd et al., 1996). High-quality friendships frequently serve as emotional and cognitive resources that help children adapt to stress and more successfully cope with social demands and interactions with classmates (Baker & Hudson, 2013). There is also evidence that measures of friendship quality that aggregate such features as cooperation, help, and conflict are related to a decrease in young children's aggression, even after accounting for initial levels of friends' and children's aggressive behaviors (Engle et al., 2011; Salvas et al., 2011). Conversely, low-quality friendships that are high in negative features such as conflict or rivalry are linked to an overall negative style of interaction that promotes disruptive behavior and poor adjustment (Engle et al., 2011; Ladd et al., 1996), unless the children involved in these conflicts have the skills to resolve them in an adaptive manner (Salvas et al., 2014; Salvas et al., 2016).

Maintaining a positive relationship with a friend may directly contribute to improved school engagement and academic competence in children, independent of experiences at the group level (Erath et al., 2008; Ladd et al., 1996). More specifically, positive friendships likely encourage children to remain motivated in school and may facilitate competent completion of academic work via shared study sessions, both of which promote school performance (Wentzel, 2009; Wentzel & Muenks, 2016). However, negative features of friendship quality may matter more than the positive features as shown by Sebanc et al. (2016). Across three time-points in the transition from elementary to early middle school, these authors found consistent bidirectional contributions to and from academic achievement but only with negative friendship quality: academic achievement negatively predicted negative friendship quality, which in turn negatively predicted academic achievement. In contrast, Vitaro and his colleagues found no contribution of best friend's relationship quality to child academic performance from kindergarten to grade 1. Importantly, this study controlled for possible genetic and shared environmental influences through the use of the MZ-twin difference method, as well as other social experiences such as relationship quality with the larger peer group (i.e., rejection and victimization), relationship quality with the teacher, and parental hostility-coercion (Vitaro et al., 2012).

In sum, features of friendship such as conflict resolution or prosocial behaviors practiced among friends may impact children's behavior directly or help explain the main effect of friendship participation or friends' behavior (i.e., operate as mediators). Other features of friendships such as the provision of companionship and social support (Birch & Ladd, 1996), or the provision of instrumental assistance and feelings of security (Wentzel, 1996), may help explain how friendship participation can impact children's emotional development. Together, these features of friendship can also impact school engagement and academic performance, although evidence of such notions is currently inconclusive. The evidence in support of a moderating role of friendship quality with respect to the contribution of friends' characteristics, notably aggression, on such outcomes is also inconclusive. In contrast, the literature showing a moderating (i.e., buffering) role of friendship quality on the stability of personal dispositions toward anxiety (Baker & Hudson, 2013) or with respect to the psychosocial correlates often associated with anxiety, such as loneliness feelings and low self-efficacy, are more consistent (Erath et al., 2010).

Peer Processes at the Group Level

A second field of research on the influence of peers on children's socioemotional and academic developmental outcomes is focused on experiences within the larger peer group, such as the classroom. With the transition to elementary school, children move into classrooms with 20–30 age-matched peers, with whom they need to function for a significant time during the day. The resulting social interactions between peers within a classroom may result in children developing appreciation for specific classmates, which may result in dyadic friendships (as discussed above). However, apart from dyadic appreciation, these group interactions may also result in evaluations of each child as well as of social behaviors in general by the group or classroom. In the following section, we will discuss two different peer processes within the larger peer group that may contribute to socioemotional and academic development in elementary school children. The first process refers to the evaluation or appreciation of individual children by the classroom peers. This evaluation may result in different degrees of social preference, or social status, of children in the classroom, with some being liked and others disliked by their classroom peers. In light of this, we will also discuss peer victimization as one of the processes associated with being disliked by classroom peers. The second process is the evaluation or appreciation of behaviors in the classroom, which we call peer social norms. This may refer to the presence of unwritten rules or norms about what behavior the classroom as a whole sees as appropriate or normative.

Peer Evaluation of Individual Children Within the Classroom Whenever children are placed in a group, or classroom, children within this classroom will start to evaluate their classmates. As a consequence of this evaluation process, children receive a "social status" within their classroom (Coie et al., 1982). That is, some children become well liked and accepted/prefered by their classmates, whereas others become disliked and rejected by their classmates. Especially a negative appreciation by peers may emerge swiftly. In their pioneering work on peer social status, Coie and Kupersmidt (1983) placed boys who were unacquainted with each other into laboratory playgroups and compared their emerging peer status in this new peer group with the status children had in their classroom. They found that within three play sessions, children who were rejected in their classroom also developed a rejected status among their new, previously unacquainted playgroup peers. A study by Gooren et al. (2011) among kindergartners confirmed that already in kindergarten some children develop a poor social status among peers, which becomes stable thereafter. Other studies covering longer developmental spans confirm that the poor social status that a child develops in kindergarten likely translates into a stable poor social position in early elementary school (van Lier & Koot, 2010) and across the elementary school years (Ladd, 2006).

Peer victimization is defined as being the recipient of repeated aggression in which one or more bullies intend to harm or disturb another person physically, verbally, or psychologically/relationally (Boulton & Underwood, 1992; Nansel et al.,

2001; Olweus, 1978; Wolke et al., 2001). Although peer victimization does not necessarily follow from poor peer appreciation, some parallels between peer rejection and peer victimization have been observed. Like peer rejection, peer victimization may emerge swiftly in elementary school and tends to become stable (van Lier et al., 2012; Pouwels et al., 2016). Importantly, crosslinks between peer social status and peer victimization have been found in that rejected children become victims of bullying, while victimized children in turn become progressively more rejected by peers (Ladd & Troop-Gordon, 2003; van Lier & Koot, 2010). As such, peer rejection may be considered a more passive evaluative process that is not necessarily expressed directly, whereas peer victimization is the behavioral enactment of rejection toward the child by at least some members of the larger peer group.

Peer Evaluation of Behaviors Within the Classroom Another peer process at the larger group level that has received considerable attention concerns the role of peer social norms for children's socioemotional and academic development (Henry et al., 2000). Peer social norms can be seen as unwritten rules within a group about what behavior seems appropriate within that group as a whole. Peer social norms derive from group consensus about what is appropriate behavior; thus, norms are shared within a group setting. Different forms of peer social norms have been described (Veenstra et al., 2018). Descriptive norms reflect what kind of behavior is most prevalent in a given group. Injunctive norms reflect what is considered appropriate in the group by asking group members to express their attitudes toward a behavior and aggregating all responses into one single attitude score that reflects the injunctive norm of the group. Norm salience captures the degree to which a behavioral norm is made salient or explicit to group members, e.g., via social rewards or sanctions. Norm salience is generally measured by the within-group correlation between behavior and social status (e.g., social acceptance, popularity or social rejection).

Impact of Peer Processes at the Group Level on Children's Socioemotional and Academic Development Many studies have documented the association of peer processes in the larger group with socioemotional development (Deater-Deckard, 2001; Hawker & Boulton, 2000; Parker et al., 2015; Rubin et al., 2006; Takizawa et al., 2014) as well as with academic development (Nakamoto & Schwartz, 2010; Véronneau et al., 2010; Vitaro et al., 2012). In line with our methodological guidelines, we highlight some studies using designs that enable studying direction of effects. The study of direction of effects seems important as one of the main reasons why children may develop poor relations with their peers is their own behavioral difficulties (Coie & Kupersmidt, 1983; Dodge, 1983; Teräsahjo & Salmivalli, 2003).

A number of longitudinal cross-lagged studies, which collectively cover the kindergarten and elementary school period, found that experiences of peer rejection predict increases in externalizing problems over and above existing problems (Ladd, 2006; Leflot et al., 2011; Sturaro et al., 2011; van Lier & Koot, 2010). Similar to peer rejection, peer victimization experiences were found to predict increases in externalizing problems even when reverse paths, from externalizing problems to victimization, were accounted for (Reijntjes et al., 2011; van Lier et al., 2012). The same results apply to internalizing outcomes with respect to peer rejection (Ladd, 2006; van Lier & Koot, 2010) and victimization (Reijntjes et al., 2010; van Lier et al., 2012). Longitudinal associations of peer rejection and victimization with problematic academic functioning have also been reported (Schwartz et al., 2005; Vaillancourt et al., 2013a).

In the last decades, interest grew in the effects of peer social norms on individual behavioral development and academic achievement. Before describing the relevant studies, however, it is important to note that individual children's behaviors affect peer social norms, especially descriptive peer social norms. Indeed, descriptive peer social norms are derived from the cumulative behaviors prevalent in a classroom and are therefore directly influenced by behaviors of individual classmates. However, studies have also shown that peer social norms in turn may influence individual behavioral development. For instance, Henry et al. (2000) showed that in classrooms where peer social norms against aggression were made salient (i.e., classrooms where aggressive behavior was not appreciated), aggressive behavior of individual students diminished over time. Other studies reported predictive effects of classroom peer norms on the development of social withdrawn behavior (Guimond et al., 2018), the socialization of aggressive behavior (Correia et al., 2019), the development of risk-taking behavior (Tieskens et al., 2019), and academic development (Chen et al., 2003) in elementary school children. Thus, similar to what is shown regarding other peer processes at the group level, such as social preference and victimization, the link between peer social norms and behavior development is likely bidirectional.

Given the impact of these peer processes at the group level on children's healthy psychosocial and academic development, numerous studies have been aimed at trying to understand the processes through which peer group processes may lead to problematic socioemotional and academic development.

Processes Through Which Peer Influences at the Group Level May Affect Socioemotional and Academic Development

A first notion that needs to be considered when trying to understand how peer processes at the group level may be associated with individual children's outcomes is the role of *cascading* experiences over the elementary school years. For instance, we previously showed that experiences of peer rejection relates to experiences of victimization in a transactional manner. That is, peer rejection may evoke experiences of victimization, with victimization in turn leading to further rejection of the child by the peer group (Teräsahjo & Salmivalli, 2003; van Lier & Koot, 2010). But this process may not stop there. Poorly appreciated children have increasing difficulties in forming friendships in the classroom (Salmivalli & Isaacs, 2005; van Lier & Koot, 2010). It is, however, not correct to assume that rejected children do not form friendships. As shown by Vitaro et al. (2007), while early elementary school disruptiveness predicts peer rejection, experiences of peer rejection among disruptive children in turn predict a drift toward forming relationships with (similarly) disruptive friends. Thus, disruptive peer-rejected children may develop friendships, but with peers who may help them refrain from disruptive behavior. In fact, affiliation with disruptive friends has been found to predict engagement in adolescent delinquency (see also Mehus et al., 2018). In another longitudinal study among elementary school children assessed from grade 3 to grade 8, it was shown that peer victimization, internalizing and externalizing symptoms, as well as academic functioning covary and interact with one another across the elementary school period and beyond (Vaillancourt et al., 2013b). Also in a study by van Lier et al. (2012), it was shown that externalizing problems predict academic underachievement and peer victimization, which in turn predict increases in internalizing and externalizing problems.

Children who are poorly appreciated by their peers may also encounter difficulties in the relationship with their teachers. As the authority figure in the classroom, teachers may have a unique role in building satisfying and safe "attachment" relations with children (Pianta & Stuhlman, 2004). Unfortunately, rejection by classmates has been found to also predict decreases in perceived support by teachers (Leflot et al., 2011), as well as rejection by teachers (Mercer & DeRosier, 2008). However, this does not necessarily mean that teachers develop negative relations with all rejected children and automatically become "part of the problem." To illustrate, in a longitudinal study, it was found that peer rejection predicted increases in internalizing problems via a reduction in positive self-concept (Spilt et al., 2014). However, this pathway was moderated by teacher support. Specifically, when teacher support was low, the pathway of rejection to internalizing problems via low self-concept was significant. In contrast, at high levels of teacher support, rejection was no longer associated with poor self-concept and subsequent internalizing problems (more on teacher support and teacher-child experiences in the next section).

Another important line of research has focused on the role of *distorted self-regulation and social cognition* as a possible link between peer influences and maladjustment. Healthy peer relations provide a context in which children learn to regulate their emotions and behaviors and to manage conflict (Asher & Rose, 1997; Bukowski, 2003; Hartup & Stevens, 1997). In addition, during interactions with peers, children learn to encode and interpret social situations, to decide on goals they want to achieve, to construct and evaluate their planned response, and to enact the desired behavior (Crick & Dodge, 1994). Peer rejection and victimization may hamper these developmental processes and distort children's interpretation of peer's intentions and hamper control processes to manage their impulses. As a result, rejected and victimized children may become more likely to engage in automatic and reflexive rather than effortful and reflective cognitive, emotional, and behavioral reactions to their environment (Schwartz, 2000). As Williams (2007) puts it, rejection or victimization "impairs individuals' ability to self-regulate, which inhibits their ability to utilize the cognitive/motivational resources that are necessary to avoid impulsive acts and to engage in hedonic sacrifice and delayed gratification" (p. 432).

In line with this notion, several studies found associations between poor social experiences with peers in elementary school and distorted social information patterns (Camodeca & Goossens, 2005) as well as maladaptive social cognitions (Rudolph et al., 2009). Associations were also found between poor peer experiences and problems with self-regulation among children (Stenseng et al., 2015; Trentacosta & Shaw, 2009). Moreover, studies found that the predictive link between peer rejection and aggression in elementary school was mediated by distorted social information processing patterns (Dodge et al., 2003; Lansford et al., 2010). In fact, the study by Lansford et al. (2010) suggested a vicious cycle between experiences of rejection, social cognitions, and aggression, such that experiences of rejection predicted distorted social cognitions, which then predicted both aggressive responses and more experiences of peer rejection. This cycle repeated itself throughout the studied period of kindergarten to third grade. In addition, it has been shown that peer social norms may interact with experiences of victimization in the development of risk-taking behavior (which is related to poor self-regulation) (Tieskens et al., 2019). Specifically, victimized children tended to show increases in risk-taking behavior only in those classrooms where the classroom norm was unfavorable toward risk-taking. This study shows that, to understand the relation between individual social experiences and behavioral development, it is important to consider the broader social context such as peer social norms.

Another set of processes that could help explain the link between negative social peer experiences and maladjustment involves *biological mechanisms*. There is growing evidence that experiences of peer rejection or peer victimization may get "under the skin" of elementary school children. Although these biological correlates may not directly explain the maladaptive outcomes found among rejected and victimized children (Deater-Deckard, 2001; Hawker & Boulton, 2000; Parker et al., 2015; Rubin et al., 2006; Takizawa et al., 2014), they may help us understand the possible pathway toward these outcomes. These biological correlates of peer rejection/victimization encompass (epi)genetic processes, stress system responses, and brain responses (see, for instance, Vaillancourt et al., 2013b). Each of these will be outlined in more detail in Chap. 4 (epigenetics), Chap. 5 (stress system), and Chap. 7 (brain responses) of this volume. However, a short overview is presented below.

Earlier in this chapter, basic principles of gene-environment interplay in peer relations were discussed with regard to friendships. These principles may also apply to peer influences at the larger group level (see also Chap. 3, this volume). For example, Brendgen et al. (2009) showed that a child's genetic risk for depression may increase the likelihood of being rejected by peers. This is an example of a gene-environment correlation (rGE). Gene-environment interactions (GxE) have also been reported with regard to peer influences at the larger group level. To illustrate, Brendgen et al. (2008a, b) showed that the link between victimization and aggression was stronger in girls with a genetic liability toward aggression compared to girls without such liability. Such GxE effects have also been found with respect to classroom norms. Specifically, children with a genetic disposition for aggressive behavior had an increased risk of being victimized by their peers, but only in classrooms where norms were unfavorable toward aggression. However, in classrooms where norms were favorable toward aggression, a genetic disposition for aggression protected the children against peer victimization (Brendgen et al., 2015). Interestingly, classroom norms may also moderate the expression of genetic liability for aggressive and non aggressive antisocial behavior in early and middle childhood (Brendgen et al., 2013b; Vitaro et al., 2015).

One specific mechanism underlying gene-environment interaction may be DNA methylation. DNA methylation is an *epigenetic process* that maintains gene activity or changes gene expression by activating or silencing the gene. Epigenetic alterations seem to function as a "biological mechanism for translating environmental signals into organismal molecular events" (Bick et al., 2012, p. 1418). Recently, it has been shown that childhood maltreatment, including exposure to peer victimization, may influence DNA methylation (Cecil et al., 2020; Mulder et al., 2020; Ouellet-Morin et al., 2013). For instance, Ouellet-Morin et al. (2013) studied DNA methylation in 28 MZ-twin pairs discordant for peer victimization experiences in elementary school. Both groups exhibited similar DNA methylation patterns prior to peer victimization, at 5 years of age. However, compared to their non-victim co-twins, twins who experienced peer victimization in elementary school had, on average, higher levels of DNA methylation at 10 years of age. Moreover, the higher levels of DNA methylation were correlated with lower cortisol responses to a stress task, 2 years later (at age 12 years) among the bullied twins, compared to their non-bullied co-twin. Importantly, "blunted" cortisol responses have been implicated with social and behavioral difficulties among children (Ouellet-Morin et al. 2011).

The human stress system is comprised of the autonomic nervous system (ANS) and the hypothalamic-pituitary-adrenal axis (HPA axis). Heart rate and heart rate variability are used as a proxy to measure activity of the ANS system. The primary end product of the HPA axis in humans is cortisol, which is a common measure of the activity of the HPA axis. Both hyper- and hypoactivation of the *stress system* have been linked with psychopathology outcomes (Beauchaine et al., 2007; El-Sheikh et al., 2001; Flynn & Rudolph, 2007; McLaughlin et al., 2015). Studies focusing on the associations between peer rejection or peer victimization and heart rate variability among elementary school children are rare. However, one study found that relational victimization in the classroom predicted decreased heart rate variability in 6–10-year-old children (Michels et al., 2013). Importantly, decreased heart rate variability in children has been associated with (sub)clinical levels of internalizing symptom (Dieleman et al., 2015; Hastings et al., 2008; Monk et al., 2001). There is also evidence that overall higher levels of cortisol across the day are associated with poor appreciation by classroom peers (Behnsen et al., 2018). Similarly, peer rejection has been associated with higher levels of cortisol at school, especially among children with few friends (Peters et al., 2011).

Advances in neuroscience have resulted in several studies on the association between elementary school children's peer experiences and *brain responses*. Studies on the normative neural responses to social stressors among elementary school children found that experiences of social exclusion may be processed in the brain in a similar way as physical pain (Achterberg et al., 2018; van der Meulen et al., 2017; van der Meulen et al., 2018). Other studies compared children with a history of poor classroom peer acceptance to children with a history of positive peer acceptance. The results suggest different neural activation between these two groups of children. To illustrate, in a study by Asscheman et al. (2019), children with a history of rejection were found to show higher activity during social exclusion in brain regions implicated in social cognition and emotion regulation.

To summarize, peer processes at the group level, such as social acceptance, peer rejection and victimization, and peer social norms, have been related to problematic socioemotional and academic development in children, which most likely happens in a bidirectional way. Different pathways of how peer processes may be linked to problematic socioemotional and academic development have been proposed. Firstly, developmental cascades may arise, where problems in one social domain may reinforce problems in another social domain, eventually leading to problematic behavioral development in children. Secondly, problems with self-regulation and social cognitions may mediate the link between peer processes at the group level and childhood behavioral development. Experiences at the peer group level may become embedded in the biology of the child. Biological processes such as differential epigenetic profiles, dysregulated stress system, and hyper- or hypoactivity in brain regions implicated in social cognition and emotion regulation have been found to link peer processes at the group level and problematic socioemotional and academic development. Finally, processes already described in reference to deviancy training and modeling might also help understand how social norms may influence children's behavior.

Teacher-Child Experiences at the Dyadic Level

Next to peers, teachers are important social partners who can play a key role in children's behavioral and academic development. The importance of dyadic teacher-child relationships has been emphasized by different theoretical perspectives, such as motivational models, social support models, and attachment theory (e.g., Roorda et al., 2011; Spilt et al., 2019; Verschueren, 2015). For instance, self-determination theory (Deci & Ryan, 2000) considers the need to belong as a basic human need and the fulfilment of this need as a condition for learning and well-being. Positive and caring relationships with significant others can play an important role in the accomplishment of the need for belonging. This is in line with the general benefits model for social support, assuming that supportive relationships promote children's socio-emotional adjustment through strengthening their positive affect, self-worth, feelings of acceptance, and efficacy beliefs (Cohen & Wills, 1985; Rueger et al., 2016). Attachment theory is another relevant framework in this context (e.g., Pianta et al., 2003). Inspired by parent-child relationship literature, it is argued that teachers can play a role as temporary or ad hoc attachment figures and that the affective quality of the relationship can make a difference for child development (e.g., Pianta et al., 2003; Verschueren & Koomen, 2012). Indeed, studies have shown that children can use their teacher as a "safe haven" to rely on for support and security in times of stress and as a "secure base" from which to explore the environment (e.g., Koomen & Hoeksma, 2003). In studies based on attachment theory, teacher-child relationships are often conceptualized along positive and negative affective dimensions, labelled as closeness, conflict, and dependency, respectively (Pianta, 2001). *Closeness* refers to the degree of warmth and openness of communication between child and teacher and the child using the teacher as a safe haven in times of stress. *Conflict* is concerned with the degree of discordant interactions between child and teacher. *Dependency* refers to age-inappropriate child dependency to the teacher (expressed in clingy behavior, e.g.) and the failure to use the teacher as a secure base (Pianta, 2001; Verschueren & Koomen, 2012).

Most research regarding the role of affective teacher-child relationships for students' adjustment has focused on preschoolers, or children transitioning from preschool to elementary school. Longitudinal studies in elementary school, especially in the later years, are scarcer (Pakarinen et al., 2018; Verschueren, 2015). Some authors (e.g., Verschueren, 2015) have argued that the role of teachers as a safe haven may become less prominent in middle and late childhood and that it shifts from proximity (in early childhood) to availability later on. However, these authors also emphasized the continued importance of teachers as a secure base from which to explore and the likely continued importance of teachers' safe haven function for vulnerable children (Pakarinen et al., 2018; Rucinski et al., 2018; Verschueren, 2015). The available evidence largely supports these assumptions, as summarized below.

Closeness/Support and Conflict Most research has focused on closeness or support as indicators of positive teacher-student relationships and on conflict as an indicator of negative relationships, whereas dependency has received less attention (but see further). A meta-analysis of 52 longitudinal studies demonstrated that both positive (i.e., close, supportive) and negative (i.e., conflicted) teacher-student relationships uniquely predicted later academic adjustment, i.e., student engagement and achievement, in both primary school (covering preschool, kindergarten, and elementary school) and secondary school (Roorda et al., 2017). Taking a closer look at single studies in elementary school, close and non-conflicted relationships have been shown to predict academic adjustment over time even when controlling for a large number of child characteristics and contextual characteristics (Maldonado-Carreño & Votruba-Drzal, 2011; O'Connor & McCartney, 2007; Spilt et al., 2012a).

For socioemotional outcomes, a number of studies demonstrated that teacher-student *conflict* predicts externalizing behavior within and across elementary school years (e.g., Buyse et al., 2009; Ettekal & Shi, 2020; Rucinski et al., 2018). For instance, Buyse et al. (2009) found that higher teacher-student conflict in first grade predicted higher levels of aggression in third grade, when controlling for a range of child and context variables. Just like studies in preschool (e.g., Doumen et al., 2008; Roorda et al., 2014), Ly and Zhou (2018) found that teacher-student conflict predicted externalizing behavior when using cross-lagged panel modelling controlling for the reverse effect, i.e., of behavior on conflict. However, two other studies in elementary school, one with an early elementary sample (Skalická et al., 2015) and the other with a late elementary school sample (Pakarinen et al., 2018), only found effects of students' externalizing behavior on conflict with the teacher. It is possible that, by elementary school, students' externalizing behavior and/or their behavioral reputation become more stable and thus less likely to be influenced by teachers. Although further discussion is beyond the scope of this section, these findings illustrate the importance of carefully controlling for potentially confounding effects in longitudinal research, and future research including both effects, i.e., of teacher-student relationships on externalizing behavior and vice versa, is needed. Findings regarding the effect of conflict on internalizing behavior are mixed: a recent study in grades 3–5 found conflict to positively predict these problems over time (e.g., Rucinski et al., 2018), while other studies comprising either younger or older elementary students did not (e.g., Ly & Zhou, 2018; O'Connor et al., 2012; Pakarinen et al., 2018). As for the effects of *closeness/support*, Maldonado-Carreño and Votruba-Drzal (2011) found that increases in teacher-student relationship quality were associated with reductions in internalizing and externalizing student behavior throughout elementary school. In addition, a recent study by Ettekal and Shi (2020) using a person-centered approach demonstrated that low teacher-student warmth in grade 1 co-occurred with more conduct problems which persisted over time until grade 12. Other longitudinal studies, however, have not found consistent evidence for main effects of *closeness* in elementary school. Although a meta-analysis of mainly cross-sectional studies demonstrated that more positive teacher-student relationships were associated with less externalizing child behavior (Lei et al., 2016), several longitudinal studies did not find closeness to predict less externalizing behavior over time in early or late elementary school (Buyse et al., 2009; Ly & Zhou, 2018; O'Connor et al., 2012; Pakarinen et al., 2018; Rucinski et al., 2018). Regarding internalizing behavior, a cross-lagged study by Spilt et al. (2019) found that a supportive teacher-student relationship predicted lower depressive symptoms across grades 2 and 3, but neither Ly and Zhou (2016) nor Rucinski et al. (2018) found a similar effect. Some of these findings are in line with Verschueren's (2015) assumption that, in general, the safe haven function of the teacher may become somewhat less prominent in middle childhood. However, in line with Verschueren's hypothesizing, there is considerable evidence regarding the moderating role of closeness for children at risk, which will be discussed later

As mentioned earlier, research regarding the role of *dependency* as a teacher-student relationship dimension is scarce. A recent meta-analysis by Roorda et al. (2021) identified 28 studies involving 7849 children from preschool to upper elementary. In the subsample of 14 longitudinal studies, it was demonstrated that dependency negatively predicted later student engagement, achievement, and prosocial behavior and positively predicted internalizing and externalizing behavior. For internalizing behavior, the effects of dependency were larger in the higher grades. Most of the studies reported in this meta-analysis did not control for initial levels of the outcome variables. The few that did also found evidence for dependency as a negative relationship dimension. For instance, a study in elementary school, controlling for initial victimization, sex, and teacher-student closeness and conflict, found that higher dependency increased the risk of both physical and relational victimization later on (Troop-Gordon & Kopp, 2011). Zee et al. (2013), controlling for a number of child characteristics, found that dependency negatively predicted students' future motivational beliefs. Although more longitudinal research is needed, the existing evidence regarding the detrimental effects of dependency suggests that teachers maintain their role as a secure base for elementary school students (cf. supra, Verschueren, 2015). Of note, all these studies, including those involved in the meta-analysis, were conducted in countries with a rather individualistic orientation. Scholars have suggested that dependency may have a different, more positive meaning in collectivistic cultures, especially for young children (e.g., Gregoriadis & Grammatikopoulos, 2014; Gregoriadis et al., 2021), but longitudinal research is lacking to our knowledge.

All single studies presented so far have longitudinal designs. Experimental studies regarding the role of teacher-student relationship quality are scarce. A few studies found positive effects of interventions aimed at improving the relationship of teachers with at-risk preschoolers on students' externalizing behaviors (e.g., Vancraeyveldt et al., 2015; Williford et al., 2017) and on their behavioral engagement and achievement (Van Craeyevelt et al., 2017). Recently, Duong et al. (2019) found that a brief teacher training aimed at promoting positive relationships with their students in middle school resulted in improvements in the relationship, behavioral engagement, and decreases in disruptive behavior. Yet, no studies in elementary school are known to us. Similarly, to the best of our knowledge, main effects of teacher-student relationship quality in elementary school have not been investigated in genetically informed studies. One study with monozygotic twins investigated the effects of social experiences in kindergarten and found that within-pair differences in teacher-child relationship quality predicted differences in academic achievement in first grade (Vitaro et al., 2012).

Teacher-student relationships have been investigated not only as main effects but also as moderators. According to the academic risk hypothesis, the social environment of the classroom, i.e., the teacher-student relationship quality, will particularly affect vulnerable students, as they have the most to gain or to lose (Hamre & Pianta, 2001; Roorda et al., 2011). As mentioned earlier, teacher-student closeness has been shown to protect students at risk for developing academic or socioemotional difficulties, thereby suggesting that teachers maintain their role as a safe haven for at-risk children. More specifically, longitudinal studies have demonstrated that a close and supportive relationship with the teacher can play a protective role for students' socioemotional and academic adjustment in case of risk for school difficulties caused by internalizing or externalizing behavioral problems, academic problems, negative caregiving experiences, or demographic risks such as minority status and low maternal education. In addition, there is evidence that high conflict and high dependency may exacerbate the negative effects of student risks on academic and socioemotional outcomes (for an overview, including studies in elementary school, see Sabol & Pianta, 2012). In an experimental study, Vandenbroucke et al. (2018) demonstrated the protective role of a supportive teacher-student relationship for students' working memory, an executive function important in learning. For students reporting a negative relationship with their parents, a supportive message from their teacher resulted in better working memory performance in an induced social stress situation; this effect of teacher support was not found for students with positive parent-child relationships exposed to the same stress. In addition, the only genetically informed, i.e., twin, study regarding this topic revealed that a positive teacher-student relationship buffered the genetically mediated expression of aggression among grade 1 children (Brendgen et al., 2011).

Regarding the role of age, it has been demonstrated that teacher-student relationship quality decreases throughout elementary school. For instance, Collins et al. (2017) found that teacher-student closeness decreased and conflict increased from grade 1 to grade 6. The studies presented above, however, do not provide clear evidence that the relationship would be less influential for late than for early elementary students. In their 2011 meta-analysis, including both cross-sectional and longitudinal studies, Roorda and colleagues even found stronger links between teacher-student relationship quality and academic engagement and achievement for older than for younger students. As for gender, it has been found that teachers, in general, report higher-quality relationships with girls than with boys (McGrath & van Bergen, 2015, for an overview). This may be explained by, among others, girls' higher attunement to teachers' behavioral expectations and girls' higher openness to teachers and/or to a gender "match," as most teachers are female as well (McGrath & van Bergen, 2015; Spilt et al., 2012b). Scholars have argued that lower-quality relationships put boys as a group at higher academic risk and may partly explain boys' lower average engagement and achievement and higher disruptive behaviors (McGrath & van Bergen, 2015; Roorda et al., 2014). Moreover, a few studies demonstrated that teachers also have a differential impact on both genders, but findings are not consistent. In Roorda et al.'s (2011) meta-analysis, both positive and negative teacher-student relationships were linked more strongly to boys' engagement, whereas positive relationships showed stronger correlations with girls' achievement. Baker (2006) found that girls benefited more than boys from teacher-child closeness in terms of socioemotional outcomes. A similar result was found in a study in kindergarten, which at the same time showed a stronger negative effect of conflict on the behavior of boys, as compared to girls (Hamre & Pianta, 2001).

A number of mechanisms have been proposed to account for teacher-student relationship effects on student outcomes (e.g., Verschueren, 2015; Verschueren & Koomen, 2012), but empirical studies that tested for them are limited. First, referring to attachment theory, it has been suggested that positive and negative relationships promote and undermine, respectively, children's stress regulation, which may, in turn, affect their academic and behavioral adjustment. This mechanism may partly explain less conclusive effects of teacher-student relationship quality in elementary compared to preschool, i.e., younger children are more dependent on their caregivers because their stress regulation capacities are less developed (Verschueren & Koomen, 2012). Of note, Ahnert et al. (2012) did find evidence for the effects of teacher-student relationship quality on first grade students' stress regulation, as expressed by their cortisol levels. Second, referring to attachment theory, Doumen et al. (2011) hypothesized that children who have positive relationships with their caregivers, such as their teachers, may consider themselves as more worthy, whereas negative relationships may give children the message that they are less valuable. Indeed, it was demonstrated that higher teacher-student conflict in first grade predicted later student self-esteem which, in turn, affected their behavioral development. Further, Roorda et al. (2017) demonstrated in their meta-analysis that the effects of positive and negative teacher-student relationships on academic achievement were partially mediated by academic engagement in primary as well as secondary school. This may be explained by motivational theories, i.e., students may become more engaged for school when their basic needs for belonging, competence, and structure are fulfilled through positive relationships with their teachers and, in turn, perform better (e.g., Deci & Ryan, 2000). Another explanation may be provided by attachment theory: When students have positive relationships with their teachers, the resulting emotional security may facilitate their engagement in class, which in turn promotes their achievement (Roorda et al., 2017).

In sum, there is meta-analytic evidence for the longitudinal effects of both positive and negative teacher-student relationships on academic engagement and achievement in elementary school. Less longitudinal studies have been conducted regarding the role of these relationships in students' socioemotional development, and findings are somewhat mixed. However, several studies provide evidence for negative effects of conflict and dependency and for positive effects of supportive and close relationships on student behavioral adjustment, in particular for at-risk children. These findings support the assumption that teachers maintain their role as attachment figures in elementary school, i.e., their role as a secure base and a safe haven, the latter mainly for at-risk children. Future research may try to replicate these findings, unravel inconsistent findings, investigate the effects of teacher-student relationships on more positive socioemotional outcomes (such as self-esteem and prosocial behavior), control for student effects on the relationship whenever possible, use more experimental and genetically informed designs, and further investigate *how* teacher-student relationships shape student development in elementary school.

Teacher-Child Interactions at the Group Level

Teachers do not only relate to individual students, but they also interact with their class as a group. According to the Teaching through Interactions framework (Hamre et al., 2014; Hamre et al., 2013), these classroom interactions are an important vehicle for learning. In their study based on this framework in pre- and elementary school, Hamre et al. (2013) identified and empirically validated three distinct domains of class-level interactions, i.e., emotional support, instructional support, and classroom organization and management. *Emotional support* refers to the promotion of students' socioemotional adjustment through warm and sensitive interactions with the class group and is related to the dimension of closeness and support in the previous section. *Instructional support* concerns the use of interactive strategies to provide students with high-quality learning opportunities, such as scaffolding and linking new information to their existing knowledge base. Finally, *classroom organization and management* refers to the way in which the classroom is arranged and organized in order to facilitate appropriate student behavior and learning (Hamre et al., 2013). Like for dyadic teacher-student interactions, most research regarding class-level interactions has been conducted in preschool. However, there is longitudinal evidence for the effects of classroom-level teacher-student interactions on academic and socioemotional development in elementary school as well. Regarding academic adjustment, for instance, Vernon-Feagans and colleagues (2019) demonstrated that children who experienced better instructional and emotional support and classroom organization from kindergarten throughout third grade had higher third grade literacy scores, even when controlling for child and family background variables and teachers' specific literacy instruction. Pianta et al. (2008) found that emotional and instructional support quality predicted both reading and math achievement trajectories from preschool to fifth grade. In addition, a cross-lagged study by Pakarinen et al. (2014) showed that lower teacher emotional support toward the class group in first grade predicted higher levels of avoidance behaviors in academically challenging situations in second grade, which is consistent with attachment theory and self-determination theory. Finally, regarding externalizing outcomes, a study in fifth and sixth grade revealed that lower teacher emotional support in fall predicted higher levels of student disruptive behavior in spring of the same school year (Shin & Ryan, 2017).

The importance of classroom-level teacher-student interactions has been demonstrated in intervention studies as well. A meta-analysis of randomized and non-randomized studies by Korpershoek et al. (2016) revealed positive effects of interventions aimed at promoting one or more of the abovementioned dimensions (i.e., emotional support, instructional support, classroom management) on elementary students' academic, behavioral, and socioemotional (but not motivational) outcomes. In particular, the dimension of classroom organization and management has been the focus of many interventions. Often, these interventions are at least partly based on learning theory (e.g., Cowan & Sheridan, 2009) and aim at strengthening teachers' skills to shape the antecedents (e.g., proactively stating clear rules and installing routines) and consequences (e.g., reinforcement) of students' behavior, in order to promote desirable and reduce disruptive behaviors. Randomized controlled trials with behavioral interventions have yielded consistent evidence for the effectiveness of this approach. A meta-analysis identified 19 randomized controlled trial (RCT) studies which tested the effects of a behavioral classroom-based intervention, either in itself or as part of a larger intervention (Veenman et al., 2018). Overall, these interventions reduced teacher-rated disruptive (i.e., hyperactive and oppositional) behavior and increased observed on-task behavior. In addition, single RCTs have found beneficial effects of the Good Behavior Game, a classroom behavior management program for teachers, on students' internalizing problems as well (e.g., Kellam et al., 2008; Van Lier et al., 2005). When finding positive effects in RCTs, researchers usually presume that these effects are driven by the theoretical mechanisms underlying the intervention, in this case teacher's classroom management, and do not further investigate these mechanisms (McKinnon, 1994; Saarento et al.,

2015). In an effort to fill this gap, an RCT investigated whether the effects of the Good Behavior Game indeed were mediated by improvements in teachers' classroom management skills (Leflot et al., 2010; Spilt et al., 2016). It was demonstrated that, compared to control teachers, Good Behavior Game teachers' compliments increased and their reprimands decreased and that these changes in teacher behavior mediated the effect of the intervention on students' disruptive and on-task behaviors.

Similar to dyadic interactions and in line with a moderation perspective, teacher-student interactions at the group level seem to matter most for students at risk (Rucinski et al., 2018). For instance, Hamre and Pianta (2005) found that in emotionally and instructionally supportive classrooms, high-risk (based on demographic characteristics and individual vulnerabilities) students' achievement scores and quality of the relationship with their teacher were similar to low-risk students, whereas high-risk students in low supportive classrooms had lower achievement and more conflict with their teacher than low-risk students. Emotional support at the classroom level was also found to buffer the risk of later internalizing problems in elementary students (Griggs et al., 2016). In addition, randomized controlled trials with interventions promoting teachers' classroom management, such as the Good Behavior Game, found larger effects on student behavior for children with high levels of disruptive behavior or other risks at baseline (e.g., Flower et al., 2014; Leflot et al., 2013). However, this was not found consistently. In their meta-analysis, Veenman et al. (2018) did not find moderation by severity of initial problem behavior. Positive effects of behavioral interventions on students' adjustment were not dependent on students' gender or age either.

Based on theory and research, a number of explanatory mechanisms of teachers' interactions with their class group can be put forward. First, findings by Veenman et al. (2018) and Leflot et al. (2010) support learning theory, i.e., that teachers can shape their students' learning and social behaviors by proactively installing clear rules and routines (antecedents) and by reinforcing desirable behaviors (consequences). Second, according to social learning theory (Bandura, 1977), students learn from the observation of the behavior of others and from the consequences of these behaviors. Hence, teachers can be considered as role models for their students. When teachers interact friendly and supportively with their students, they model appropriate relational skills and increase the likelihood that children will behave nicely and be supportive. Coercive or even hostile teacher-student interactions model negative relational patterns and may, thereby, provoke or enhance power-assertive student behavior (e.g., Serdiouk et al., 2015). In line with this theorizing, Weyns et al. (2017) showed that teachers' observed praise and reprimands at the beginning of second grade predicted the development of students' relational aggression from second to fourth grade: Whereas relational aggression generally increased during that period, it increased at a slower or a faster rate when teachers displayed, respectively, more praise or more reprimands (Weyns et al., 2017). In addition, like for dyadic teacher-student relationships, attachment theory and motivational theories may provide explanatory frameworks. In line with the first, Ahnert et al. (2012) demonstrated that, besides dyadic interactions, classroom-level teacher-student interactions predicted students' cortisol profiles: Compared to high-supportive classrooms, students in low-supportive classrooms displayed flatter cortisol profiles, suggesting they were less able to downsize their stress. This finding further supports the idea that teachers affect (young) students' adjustment by influencing their stress regulation capacities. Studies based on self-determination theory often focus on adolescence and have demonstrated that teacher-student interactions at the classroom level impact student development through motivational processes. For instance, it has been found that psychologically controlling teaching predicted lower academic and socioemotional adjustment and this effect was mediated by need frustration (e.g., Filippello et al., 2019). Taken together, these findings suggest that teacher-student interactions at the classroom level remain important when children grow older; yet the relative importance of certain types of interactions and of the underlying mechanisms may change, i.e., shift from supporting students' self-regulation to promoting student motivation through autonomy supporting interactions.

In sum, longitudinal research has found effects of both dyadic and classroom-level teacher-student interactions on students' academic and socioemotional development. But to which extent do both types of interactions have distinct effects on students? Longitudinal research including both dyadic and classroom-level teacher-student interactions in the prediction of student outcomes is limited but emergent. A few studies in preschool have found that dyadic and class-level interactions uniquely contribute to subsequent student adjustment (e.g., Cadima et al., 2016; Weyns et al., 2019). A study in grades 3–5 found that dyadic teacher-student relationship quality, but not group-level interactions, predicted changes in student internalizing and externalizing behavior (Rucinski et al., 2018). According to the authors, the nonsignificant contribution of the group-level variable may be due to the low number of classrooms in their sample, so further research is needed to uncover this issue. In addition, a number of studies have investigated the interplay between dyadic and classroom teacher-student interactions in their link with student outcomes, but the findings are not conclusive (Rucinski et al., 2018). For instance, Rucinski et al. (2018) found that teachers' emotional support toward the class group mitigated the link between teacher-student conflict and increases in students' aggression. More specifically, the effect of conflict on aggression was lower in highly supportive classrooms, compared to low-supportive classrooms. This is consistent with studies in preschool demonstrating a protective role of classroom-level interactions for students with low-quality dyadic interactions with their teacher (e.g., Buyse et al., 2008). However, in Rucinski's study, class-level teacher support did not moderate other links between dyadic interaction quality and student outcomes. Neither did a study by Hughes et al. (2006) in first grade find evidence for such moderation. To conclude, until future research clarifies the unique and joint effects of dyadic and group-level teacher-student interactions on children's development, it seems premature to assume that teachers' interactions with their class group can mitigate or compensate for low-quality dyadic relationships (Rucinski et al., 2018). Hence, within the current state of the art, it is important to train pre- and in-service elementary school teachers not only to manage their classrooms and interact supportively with the class group but also to build high-quality dyadic relationship with their students, in particular their students at risk.

Interplay Between Peer and Teacher Experiences

As could be seen in the previous sections, several studies examined the additive or interactive role of different types of peer or teacher experiences while controlling for other types of experiences. This should become standard procedure given the overlap between different types of peer or teacher experiences. However, few studies examined the possible interplay among the peer and/or teacher experiences over time, but there are some exceptions. To illustrate, van Lier and Koot (2010) found that peer rejection and peer victimization, but not friendship participation, are transactionally (i.e., bidirectionally) related from kindergarten to grade 4. These authors also found that peer rejection and peer victimization link kindergarten externalizing problems with fourth grade internalizing problems, similarly for boys and girls. However, only peer rejection, not victimization nor friendship participation, contributed to the development of externalizing problems. In another study, van Lier and his colleagues investigated the role of peer rejection and best friend's externalizing behavior in the development of externalizing behavior in children followed from kindergarten to grade 3 (Sturaro et al., 2011). Using a cross-lagged design, these authors found that peer rejection, but not best friend's externalizing problems, consistently positively predicted externalizing problems above and beyond prior levels of problem behavior. All findings were similar for boys and girls. These results concur with findings from Vitaro and his colleagues (2007) who also found that peer rejection, not friends' externalizing problems, played an important role during early childhood with respect to the linkage between early disruptive behaviors and later violent delinquency and substance use. However, friends' externalizing problems by late childhood played a necessary, albeit partial, mediating role in the process linking early disruptiveness and later violent delinquency, but not in the process linking early disruptiveness and later substance use (Vitaro et al., 2007). These results underline the dynamic role each peer experience can play with development. They also show the importance, if not the necessity, of considering different types of peer experiences (and different outcomes) in the same study.

Experiences with peers and experiences with teachers are also interconnected, although it is not clear whether this is because teachers and peers react in a similar way to children's characteristics or because they influence each other (Howes et al., 1994). In support with the latter view, White and her colleagues experimentally manipulated teachers' behavior and showed that verbal comments expressed publicly by the teacher toward a particular child influenced classmates' perception of the child, independent of the child's actual behavior (White & Jones, 2000; White et al., 1996). Accordingly, longitudinal research demonstrated that teacher behavior toward students predicted peer perceptions of teachers' disliking 3 months later and this, in turn, predicted peers' disliking of those students 6 months later (Hendrickx et al., 2017). Going one step further, Sette and her colleagues showed that teachers' liking of students predicted peer inclusion which, in turn, accounted for the link between teachers' liking and changes in academic achievement from grade 5 to grade 6 (Sette et al., 2020). The influence between the teacher and the peer group may, however, be reciprocal. For example, Mercer and DeRosier (2008) found that peer rejection and teacher preference (i.e., how much a teacher likes a child) reciprocally predict each other over four data points spanning from fall of grade 3 to spring of grade 4. This result was obtained despite controlling for concurrent children's aggression. Other studies also found transactional links between peer relationships (peer liking or peer-perceived popularity) and teacher-child relationships (i.e., support and conflict) throughout childhood, net of children's behavioral characteristics (De Laet et al., 2014; Hughes & Chen, 2011). Taken together, these findings suggest that not only peers' but also teachers' liking or disliking of a student is influenced by their perception of how much that student is liked by others and thus act as social referents for each other. However, reciprocal links have not always been found. In some studies with children aged between 8 and 12 years old, peer acceptance predicted changes in teacher support, but teacher support did not predict changes in peer acceptance (Leflot et al., 2011; Weyns et al., 2018). Yet, in another study involving kindergarten children, peer rejection did not predict teacher preference (Taylor, 1989). Finally, in a recent study involving upper elementary school children, Demol et al. (2020) found no cross-lagged links between teacher support and peer rejection during the course of one school year when taking into account other peer experiences such as peer victimization. Interestingly, however, these authors found bidirectional (negative) longitudinal links between teacher support and peer victimization during the first part of the school year and a unilateral (negative) longitudinal link between teacher support and peer victimization during the second part, net of peer rejection. The only possible conclusion at this point is that further research regarding the dynamic interplay between different types of peer experiences and different dimensions of the teacher-child relationship is needed.

The above studies assessed the interplay between peer experiences and teacher experiences during childhood. Some also tested for the contribution of each type of social experience while controlling their overlap with the other type, as well as their overlap with children's characteristics who often precede these social experiences. However, the studies that simultaneously and repeatedly examined the contribution of peer and teacher experiences to children's behavior or school-related outcomes using a cross-lagged design spanning over two or more data points are scarce. Moreover, their findings do not always concord. For example, Mercer and DeRosier (2008) found that aggression remained relatively unaffected by either peer rejection or teacher preference during late childhood. In contrast, Leflot et al. (2011) found that peer preference, but not teacher support, contributed to changes in externalizing problems during the middle childhood period. In turn, Weyns et al. (2019) found that interactions with the teacher, both at the individual and at the classroom level, contributed to changes in internalizing and externalizing problems in kindergarteners, but peer preference did not. On the other hand, it seems that both peer rejection and teacher-child relationship quality are important when the outcome is academic performance in young children, even after accounting for genetic effects (Vitaro et al., 2012). In accordance with this last comment, De Laet et al. (2015), as well as Weyns et al. (2018) found additive and unique effects of teacher-student relationships and peer status (likeability and popularity) on school engagement in early and late elementary school children. Together, these results underline the possibility that experiences with peers and experiences with teachers might play different, albeit complementary, roles depending on the nature of the outcome (i.e., socioemotional, behavioral, or academic) or on children's age. This, however, remains largely speculative at the moment.

Conclusions and Future Directions

There are a number of points that can be derived from the present chapter: First, we need to consider several peer and teacher experiences simultaneously to control for their overlap and asses their unique role, but also to examine their possible interplay over time and their possible interactive (vs additive) contributions to children' development. Second, we need to consider different outcomes over several data points, possibly within the same school year, as the role of different social experiences may differ depending on the type of outcome considered and children's developmental status or degree of acquaintance with peers or teacher. Third, we need to examine possible behavioral, cognitive, affective, neuro-physiological, or epigenetic mechanisms that can account for different types of social experiences during the elementary school years as well as different factors biological, behavioral, cognitive or social that might mitigate or aggravate their contribution. Finally, we need to consider teacher and peer experiences as potential moderators of the link between children's characteristics and different outcomes, although they may not (always) produce main effects (see, e.g., Vitaro et al., 2018, in reference to aggression).

In parallel to our efforts to integrate different social experiences, we need to consider strong methodologies, including more experimental studies. These experimental studies may take the form of intervention/prevention efforts to improve experiences with peers and experiences with the teacher with the goal of increasing children's social behavior, well-being, and academic performance or engagement (e.g., Pahigiannis & Glos, 2020; Vancraeyveldt et al., 2015). In turn, intervention/prevention studies can help pinpoint the causal role of these social experiences.

References

Achterberg, M., van Duijvenvoorde, A. C. K., van der Meulen, M., Bakermans-Kranenburg, M. J., & Crone, E. A. (2018). Heritability of aggression following social evaluation in middle childhood: An fMRI study. *Human Brain Mapping, 39*(7), 2828–2841. https://doi.org/10.1002/hbm.24043

Adams, R. E., & Bukowski, W. M. (2007). Relationships with mothers and peers moderate the association between childhood sexual abuse and anxiety disorders. *Child Abuse & Neglect, 31*(6), 645–656. https://doi.org/10.1016/j.chiabu.2006.12.011

Adams, R. E., Santo, J. B., & Bukowski, W. M. (2011). The presence of a best friend buffers the effects of negative experiences. *Developmental Psychology, 47*(6), 1786–1791. https://doi.org/10.1037/a0025401

Ahnert, L., Harwardt-Heinecke, E., Kappler, G., Eckstein-Madry, T., & Milatz, A. (2012). Student-teacher relationships and classroom climate in first grade: How do they relate to students' stress regulation? *Attachment & Human Development, 14*, 249–263. https://doi.org/10.1080/14616734.2012.673277

Archer, J., & Lloyd, B. (2002). *Sex and gender* (2nd ed.). Cambridge University Press. https://doi.org/10.1017/CBO9781139051910

Asher, S. R., & Rose, A. J. (1997). Promoting children's social-emotional adjustment with peers. In P. Salovey & D. J. Sluyter (Eds.), *Emotional development and emotional intelligence: Educational implications (p.196-230).* Basic Books.

Asscheman, J. S., Koot, S., Ma, I., Buil, J. M., Krabbendam, L., Cillessen, A. H., & van Lier, P. A. (2019). Heightened neural sensitivity to social exclusion in boys with a history of low peer preference during primary school. *Developmental Cognitive Neuroscience, 38*, 100673. https://doi.org/10.1016/j.dcn.2019.100673

Bagwell, C. L., & Coie, J. D. (2004). The best friendships of aggressive boys: Relationship quality, conflict management, and rule-breaking behavior. *Journal of Experimental Child Psychology, 88*, 5–24. https://doi.org/10.1016/j.jecp.2003.11.004

Bagwell, C. L., Newcomb, A. F., & Bukowski, W. M. (1998). Preadolescent friendship and peer rejection as predictors of adult adjustment. *Child Development, 69*, 140–153. https://doi.org/10.1111/j.1467-8624.1998.tb06139.x

Baker, J. A. (2006). Contributions of teacher–child relationships to positive school adjustment during elementary school. *Journal of School Psychology, 44*, 211–229. https://doi.org/10.1016/j.jsp.2006.02.002

Baker, J. R., & Hudson, J. L. (2013). Friendship quality predicts treatment outcome in children with anxiety disorders. *Behaviour Research and Therapy, 51*(1), 31–36. https://doi.org/10.1016/j.brat.2012.10.005

Bandura, A. (1977). *Social learning theory.* Prentice Hall. https://doi.org/10.1177/105960117700200317

Barry, C. M., & Wentzel, K. R. (2006). Friend influence on prosocial behaviour: The role of motivational factors and friendship characteristics. *Developmental Psychology, 42*(1), 153–163. https://doi.org/10.1037/0012-1649.42.1.153

Beauchaine, T. P., Gatzke-Kopp, L., & Mead, H. K. (2007). Polyvagal theory and developmental psychopathology: Emotion dysregulation and conduct problems from preschool to adolescence. *Biological psychology, 74*(2), 174–184. https://doi.org/10.1016/j.biopsycho.2005.08.008

Behnsen, P., Buil, M., Koot, S., Huizink, A., & van Lier, P. (2018). Classroom social experiences in early elementary school relate to diurnal cortisol levels. *Psychoneuroendocrinology, 87*, 1–8. https://doi.org/10.1016/j.psyneuen.2017.09.025

Berndt, T. J. (1999). Friends' influence on students' adjustment to school. *Educational Psychologist, 34*, 15–28. https://doi.org/10.1207/s15326985ep3401_2

Betts, L. R., & Rotenberg, K. J. (2007). Trustworthiness, friendships and self-control: factors that contribute to young children's school adjustment. *Infant and Child Development, 16*(5), 491–508. https://doi.org/10.1002/icd.518

Bick, J., Naumova, O., Hunter, S., Barbot, B., Lee, M., Luthar, S. S., … Grigorenko, E. L. (2012). Childhood adversity and DNA methylation of genes involved in the hypothalamus–pituitary–adrenal axis and immune system: Whole-genome and candidate-gene associations. *Development and Psychopathology, 24*(4), 1417–1425. https://doi.org/10.1017/S0954579412000806

Birch, S. H., & Ladd, G. W. (1996). Interpersonal relationships in the school environment and children's early school adjustment: The role of teachers and peers. In J. Juvonen & K. R. Wentzel (Eds.), *Social motivation: Understanding children's school adjustment* (pp. 199–225). Cambridge University Press. https://doi.org/10.1017/CBO9780511571190.011

Boivin, M., & Vitaro, F. (1995). The impact of peer relationships on aggression in childhood: Inhibition through coercion or promotion through peer support. In J. McCord (Ed.), *Coercion and punishment in long-term perspectives* (pp. 183–197). Cambridge University Press. https://doi.org/10.1017/CBO9780511527906.011

Boivin, M., Hymel, S., & Hodges, E. (2001). Toward a process view of peer rejection and harassment. In J. Juvonen & S. Graham (Eds.), *Peer harassment in school: The plight of the vulnerable and victimized children* (pp. 2265–2289). Guilford Press.

Boulton, M. J., & Underwood, K. (1992). Bully/victim problems among middle school children. *British Journal of Educational Psychology, 62*(1), 73–87. https://doi.org/10.1111/j.2044-8279.1992.tb01000.x

Brendgen, M., Boivin, M., Vitaro, F., Bukowski, W. M., Dionne, G., Tremblay, R. E., & Pérusse, D. (2008a). Linkages between children's and their friends' social and physical aggression: Evidence for a gene-environment interaction. *Child Development, 79*(1), 13–29. https://doi.org/10.1111/j.1467-8624.2007.01108.x

Brendgen, M., Boivin, M., Vitaro, F., Girard, A., Dionne, G., & Pérusse, D. (2008b). Gene–environment interaction between peer victimization and child aggression. *Development and psychopathology, 20*(2), 455–471. https://doi.org/10.1017/S0954579408000229

Brendgen, M., Vitaro, F., Boivin, M., Girard, A., Bukowski, W. M., Dionne, G., Tremblay, R. E., & Pérusse, D. (2009). Gene-environment interplay between peer rejection and depressive behavior in children. *Journal of Child Psychology and Psychiatry, 50*(8), 1009–1017. https://doi.org/10.1111/j.1469-7610.2009.02052.x

Brendgen, M., Baker, E. D., Boivin, M., Dionne, G., Vitaro, F., Girard, A., Tremblay, R., & Pérusse, D. (2011). Gene-environment processes linking aggression, peer victimization, and the teacher-child relationship. *Child Development, 82*, 2021–2036. https://doi.org/10.1111/j.1467-8624.2011.01644.x

Brendgen, M., Vitaro, F., & Girard, A. (2012). Evaluating gene-environment interplay. In B. Laursen, T. D. Little, & N. A. Card (Eds.), *Handbook of developmental research methods* (pp. 687–705). Guilford Publications.

Brendgen, M., Vitaro, F., Bukowski, W. M., Dionne, G., Tremblay, R. E., & Boivin, M. (2013a). Can friends protect genetically vulnerable children from depression? *Development and Psychopathology, 25*, 277–289. https://doi.org/10.1017/S0954579412001058

Brendgen, M., Girard, A., Vitaro, F., Dionne, G., & Boivin, M. (2013b). Do peer group norms moderate the expression of genetic risk for aggression? *Journal of Criminal Justice, 41*(5), 324–330. https://doi.org/10.1016/j.jcrimjus.2013.06.004

Brendgen, M., Girard, A., Vitaro, F., Dionne, G., & Boivin, M. (2015). Gene-environment correlation linking aggression and peer victimization: do classroom behavioral norms matter? *Journal of Abnormal Child Psychology, 43*(1), 19–31. https://doi.org/10.1007/s10802-013-9807-z

Bronfenbrenner, U., & Morris, P. A. (2006). The bioecological model of human development. *Handbook of Child Psychology*. https://doi.org/10.1002/9780470147658.chpsy0114

Buil, J. M., Van Lier, P. A., Brendgen, M. R., Koot, H. M., & Vitaro, F. (2017). Developmental pathways linking childhood temperament with antisocial behavior and substance use in adolescence: Explanatory mechanisms in the peer environment. *Journal of Personality and Social Psychology, 112*(6), 948–966. https://doi.org/10.1037/pspp0000132

Bukowski, W. M. (2003). Peer relationships. In M. H. Bornstein, L. Davidson, C. L. M. Keyes, & K. A. Moore (Eds.), *Crosscurrents in contemporary psychology. Well-being: Positive development across the life course* (pp. 221–223). Lawrence Erlbaum Associates Publishers.

Bukowski, W. M., & Hoza, B. (1989). Popularity and friendship: Issues in theory, measurement, and outcome. In T. Berndt & G. Ladd (Eds.), *Peer relationships in child development* (pp. 15–45). Wiley.

Buyse, E., Verschueren, K., Doumen, S., Van Damme, J., & Maes, F. (2008). Classroom problem behavior and teacher–child relationships in kindergarten: The moderating role of classroom climate. *Journal of School Psychology, 46*, 367–391. https://doi.org/10.1016/j.jsp.2007.06.009

Buyse, E., Verschueren, K., Verachtert, P., & Van Damme, J. (2009). Predicting school adjustment in early elementary school: Impact of teacher-child relationship quality and relational classroom climate. *Elementary School Journal, 110*, 119–141. https://doi.org/10.1086/605768

Cadima, J., Verschueren, K., Leal, T., & Guedes, C. (2016). Classroom interactions, dyadic teacher–child relationships, and self–regulation in socially disadvantaged young children. *Journal of Abnormal Child Psychology, 44*, 7–17. https://doi.org/10.1007/s10802-015-0060-5

Camodeca, M., & Goossens, F. A. (2005). Aggression, social cognitions, anger and sadness in bullies and victims. *Journal of Child Psychology and Psychiatry, 46*(2), 186–197. https://doi.org/10.1111/j.1469-7610.2004.00347.x

Cecil, C. A., Zhang, Y., & Nolte, T. (2020). Childhood maltreatment and DNA methylation: A systematic review. *Neuroscience & Biobehavioral Reviews*. https://doi.org/10.1016/j.neubiorev.2020.02.019

Chen, X., Chang, L., & He, Y. (2003). The peer group as a context: Mediating and moderating effects on relations between academic achievement and social functioning in Chinese children. *Child Development*. https://doi.org/10.1111/1467-8624.00564

Cohen, S., & Wills, T. (1985). Stress, social support, and the buffering hypothesis. *Psychological Bulletin, 98*(2), 310–357. https://doi.org/10.1037/0033-2909.98.2.310

Coie, J. D., & Kupersmidt, J. D. (1983). A behavioral analysis of emerging social status in boys' groups. *Child Development, 54*, 1400–1416. https://doi.org/10.2307/1129803

Coie, J. D., Dodge, K. A., & Coppotelli, H. (1982). Dimensions and types of social status: A cross-age perspective. *Developmental Psychology, 18*, 557–570. https://doi.org/10.1037/0012-1649.18.4.557

Collins, B. A., O'Connor, E. E., Supplee, L., & Shaw, D. S. (2017). Behavior problems in elementary school among low-income boys: The role of teacher-child relationships. *The Journal of Educational Research, 110*, 72–84. https://doi.org/10.1080/00220671.2015.1039113

Correia, S., Brendgen, M., & Vitaro, F. (2019). The role of norm salience in aggression socialization among friends: Distinctions between physical and relational aggression. *International Journal of Behavioral Development*. https://doi.org/10.1177/0165025419854133

Cowan, R. J., & Sheridan, S. M. (2009). Evidence-based approaches to working with children with disruptive behavior. In T. B. Gutkin & C. R. Reynolds (Eds.), *The handbook of school psychology* (4rd ed., pp. 569–590). John Wiley.

Crick, N. R., & Dodge, K. A. (1994). A review and reformulation of social-information processing mechanisms in children's social adjustment. *Psychological Bulletin, 115*(1), 74–101. https://doi.org/10.1037/0033-2909.115.1.74

De Laet, S., Doumen, S., Vervoort, E., Colpin, H., Van Leeuwen, K., Goossens, L., & Verschueren, K. (2014). Transactional links between teacher–child relationship quality and perceived versus sociometric popularity: A three-wave longitudinal study. *Child Development, 85*(4), 1647–1662. https://doi.org/10.1111/cdev.12216

De Laet, S., Colpin, H., Vervoort, E., Doumen, S., Van Leeuwen, K., Goossens, L., & Verschueren, K. (2015). Developmental trajectories of children's behavioral engagement in late elementary school: Both teachers and peers matter. *Developmental Psychology, 51*(9), 1292–1306. https://doi.org/10.1037/a0039478

Deater-Deckard, K. (2001). Annotation: Recent research examining the role of peer relationships in the development of psychopathology. *Journal of Child Psychology and Psychiatry, 42*(5), 565–579. https://doi.org/10.1017/S0021963001007272

Deci, E. L., & Ryan, R. M. (2000). The "What" and "Why" of goal pursuits: Human needs and the self-determination of behavior. *Psychological Inquiry, 11*, 227–268. https://doi.org/10.1207/S15327965PLI1104_01

Demol, K., Leflot, G., Verschueren, K., & Colpin, H. (2020). Revealing the transactional associations among teacher-child relationships, peer rejection and peer victimization in early adolescence. *Journal of Youth and Adolescence, 49*(11), 2311–2326. https://doi.org/10.1007/s10964-020-01269-z

Deptula, D. P., & Cohen, R. (2004). Aggressive, rejected, and delinquent children and adolescents: A comparison of their friendships. *Aggression and Violent Behavior, 9*, 75–104. https://doi.org/10.1016/S1359-1789(02)00117-9

Dieleman, G. C., Huizink, A. C., Tulen, J. H., Utens, E. M., Creemers, H. E., van der Ende, J., & Verhulst, F. C. (2015). Alterations in HPA-axis and autonomic nervous system functioning in childhood anxiety disorders point to a chronic stress hypothesis. *Psychoneuroendocrinology, 51*, 135–150. https://doi.org/10.1016/j.psyneuen.2014.09.002

Dirghangi, S., Kahn, G., Laursen, B., Brendgen, M., Vitaro, F., Dionne, G., & Boivin, M. (2015). Corumination cultivates anxiety: A genetically informed study of friend influence during early adolescence. *Developmental Psychology, 51*(4), 564–571. https://doi.org/10.1037/a0038848

Dishion, T. J., Spracklen, K. M., Andrews, D. W., & Patterson, G. R. (1996). Deviancy training in male adolescent friendships. *Behavior Therapy, 27*, 373–390. https://doi.org/10.1016/S0005-7894(96)80023-2

Dodge, K. A. (1983). Behavioral antecedents of peer social status. *Child Development, 54*(6), 1386–1399. https://doi.org/10.2307/1129802

Dodge, K. A., Lansford, J. E., Burks, V. S., Bates, J. E., Pettit, G. S., Fontaine, R., & Price, J. M. (2003). Peer rejection and social information-processing factors in the development of aggressive behavior problems in children. *Child Development, 74*(2), 374–393. https://doi.org/10.1111/1467-8624.7402004

Doumen, S., Verschueren, K., Buyse, E., Germeijs, V., Luyckx, K., & Soenens, B. (2008). Reciprocal relations between teacher–child conflict and externalizing behavior in kindergarten: A three-wave longitudinal study. *Journal of Clinical Child & Adolescent Psychology, 37*, 588–599. https://doi.org/10.1080/15374410802148079

Doumen, S., Buyse, E., Colpin, H., & Verschueren, K. (2011). Teacher-child conflict and aggressive behaviour in first grade: The intervening role of children's self-esteem. *Infant and Child Development, 20*, 449–465. https://doi.org/10.1002/icd.725

Duong, M. T., Pullmann, M. D., Buntain-Ricklefs, J., Lee, K., Benjamin, K. S., Nguyen, L., & Cook, C. R. (2019). Brief teacher training improves student behavior and student-teacher relationships in middle school. *School Psychology, 34*, 212–221. https://doi.org/10.1037/spq000296

Eivers, A. R., Brendgen, M., Vitaro, F., & Borge, A. I. H. (2012). Concurrent and longitudinal links between children's and their friends' antisocial and prosocial behavior in preschool. *Early Childhood Research Quarterly, 27*, 137–146. https://doi.org/10.1016/j.ecresq.2011.05.001

El-Sheikh, M., Harger, J., & Whitson, S. M. (2001). Exposure to interparental conflict and children's adjustment and physical health: The moderating role of vagal tone. *Child development, 72*(6), 1617–1636. https://doi.org/10.1111/1467-8624.00369

Engle, J. M., McElwain, N. L., & Lasky, N. (2011). Presence and quality of kindergarten children's friendships: Concurrent and longitudinal associations with child adjustment in the early school years. *Infant and Child Development, 20*(4), 365–386. https://doi.org/10.1002/icd.706

Erath, S. A., Flanagan, K. S., & Bierman, K. L. (2008). Early adolescent school adjustment: Associations with friendship and peer victimization. *Social Development, 17*, 853–870. https://doi.org/10.1111/j.1467-9507.2008.00458.x

Erath, S. A., Flanagan, K. S., Bierman, K. L., & Tu, K. M. (2010). Friendships moderate psychosocial maladjustment in socially anxious early adolescents. *Journal of Applied Developmental Psychology, 31*(1), 15–26. https://doi.org/10.1016/j.appdev.2009.05.005

Ettekal, I., & Shi, Q. (2020). Developmental trajectories of teacher-student relationships and longitudinal associations with children's conduct problems from Grades 1 to 12. *Journal of School Psychology, 82*, 17–35. https://doi.org/10.1016/j.jsp.2020.07.004

Fergusson, D. M., Woodward, L. J., & Horwood, L. J. (1999). Childhood peer relationship problems and young people's involvement with deviant peers in adolescence. *Journal of Abnormal Child Psychology, 27*(5), 357–369. https://doi.org/10.1023/a:1021923917494

Filippello, P., Buzzai, C., Costa, S., & Sorrenti, L. (2019). School refusal and absenteeism: Perception of teacher behaviors, psychological basic needs, and academic achievement. *Frontiers in Psychology, 10*, 1471. https://doi.org/10.3389/fpsyg.2019.01471

Flower, A., McKenna, J. W., Bunuan, R. L., Muething, C. S., & Vega jr., R. (2014). Effects of the Good Behavior Game on challenging behavior in school settings. *Review of Educational Research, 84*, 546–571. https://doi.org/10.3102/0034654314536781

Flynn, M., & Rudolph, K. D. (2007). Perceptual asymmetry and youths' responses to stress: Understanding vulnerability to depression. *Cognition and Emotion, 21*(4), 773–788. https://doi.org/10.1080/02699930600824635

Gooren, E. M. J. C., van Lier, P. A. C., Stegge, H., Terwogt, M. M., & Koot, H. M. (2011). The development of conduct problems and depressive symptoms in early elementary school children: The role of peer rejection. *Journal of Clinical Child and Adolescent Psychology, 40*(2), 245–253. https://doi.org/10.1080/15374416.2011.546045

Gregoriadis, A., & Grammatikopoulos, V. (2014). Teacher–child relationship quality in early childhood education: The importance of relationship patterns. *Early Child Development and Care, 184*, 386–402. https://doi.org/10.1080/03004430.2013.790383

Gregoriadis, A., Grammatikopoulos, V., Tsigilis, N., & Verschueren, K. (2021). Teachers' and children's perceptions about their relationships: examining the construct of dependency in the Greek sociocultural context. *Attachment & Human Development, 23*(5), 556–571. https://doi.org/10.1080/14616734.2020.1751990

Griggs, M. S., Mikami, A. Y., & Rimm-Kaufman, S. E. (2016). Classroom quality and student behavior trajectories in elementary school. *Psychology in the Schools, 53*, 690–704. https://doi.org/10.1002/pits.21941

Guimond, F.-A., Brendgen, M., Correia, S., Turgeon, L., & Vitaro, F. (2018). The moderating role of peer norms in the associations of social withdrawal and aggression with peer victimization. *Developmental Psychology, 54*(8), 1519–1527. https://doi.org/10.1037/dev0000539

Hamre, B. K., & Pianta, R. C. (2001). Early teacher-child relationships and the trajectory of children's school outcomes through eighth grade. *Child Development, 72*, 625–638. https://doi.org/10.1111/1467-8624.00301

Hamre, B. K., & Pianta, R. C. (2005). Can instructional and emotional support in the first-grade classroom make a difference for children at risk of school failure? *Child Development, 76*, 949–967. https://doi.org/10.1111/j.1467-8624.2005.00889.x

Hamre, B. K., Pianta, R. C., Downer, J. T., DeCoster, J., Mashburn, A. J., Jones, S. M., … Brackett, M. A. (2013). Teaching through interactions: Testing a developmental framework of teacher effectiveness in over 4,000 classrooms. *The Elementary School Journal, 113*, 461–487. https://doi.org/10.1086/669616

Hamre, B., Hatfield, B., Pianta, R., & Jamil, F. (2014). Evidence for general and domain-specific elements of teacher-child interactions: Associations with preschool children's development. *Child Development, 85*, 1257–1274. https://doi.org/10.1111/cdev.12184

Hartup, W. W. (1996). The company they keep: Friendships and their development significance. *Child Development, 67*(1), 1–13. https://doi.org/10.1111/j.1467-8624.1996.tb01714.x

Hartup, W. W. (2005). Peer interaction: What causes what? *Journal of Abnormal Child Psychology, 33*(3), 387–394. https://doi.org/10.1007/s10802-005-3578-0

Hartup, W. W., & Stevens, N. (1997). Friendships and adaptation in the life course. *Psychological Bulletin, 121*(3), 355–370. https://doi.org/10.1037/0033-2909.121.3.355

Hastings, P. D., Nuselovici, J. N., Utendale, W. T., Coutya, J., McShane, K. E., & Sullivan, C. (2008). Applying the polyvagal theory to children's emotion regulation: Social context, socialization, and adjustment. *Biological Psychology, 79*(3), 299–306. https://doi.org/10.1016/j.biopsycho.2008.07.005

Hawker, D. S., & Boulton, M. J. (2000). Twenty years' research on peer victimization and psychosocial maladjustment: A meta-analytic review of cross-sectional studies. *Journal of Child Psychology and Psychiatry, 41*(4), 441–455. https://doi.org/10.1111/1469-7610.00629

Hendrickx, M. M. H., Mainhard, T., Oudman, S., Boor-Klip, H. J., & Brekelmans, M. (2017). Teacher behavior and peer liking and disliking: The teacher as a social referent for peer status. *Journal of Educational Psychology, 109*(4), 546–558. https://doi.org/10.1037/edu0000157

Henry, D., Guerra, N., Huesmann, R., Tolan, P., VanAcker, R., & Eron, L. (2000). Normative influences on aggression in urban elementary school classrooms. *American Journal of Community Psychology, 28*(1), 59–81. https://doi.org/10.1023/A:1005142429725

Hodges, E. V. E., Boivin, M., Vitaro, F., & Bukowski, W. M. (1999). The power of friendship: Protecting against an escalating cycle of peer victimization. *Developmental Psychology, 35*(1), 94–101. https://doi.org/10.1037//0012-1649.35.1.94

Howes, C., Hamilton, C. E., & Matheson, C. C. (1994). Children's relationships with peers: Differential associations with aspects of the teacher-child relationship. *Child Development, 65*(1), 253–263. https://doi.org/10.1111/j.1467-8624.1994.tb00748.x

Hughes, J. N., & Chen, Q. (2011). Reciprocal effects of student–teacher and student–peer relatedness: Effects on academic self-efficacy. *Journal of Applied Developmental Psychology, 32*(5), 278–287. https://doi.org/10.1016/j.appdev.2010.03.005

Hughes, J. N., Zhang, D., & Hill, C. R. (2006). Peer assessments of normative and individual teacher-student support predict social acceptance and engagement among low-achieving children. *Journal of School Psychology, 43*, 447–463. https://doi.org/10.1016/j.jsp.2005.10.002

Kellam, S. G., Brown, C. H., Poduska, J., Ialongo, N., Wang, W., Toyinbo, P., Petras, H., … Wilcox, H. C. (2008). Effects of a Universal Classroom Behavior Management Program in First and Second Grades on Young Adult Behavioral, Psychiatric, and Social Outcomes. *Drug and Alcohol Dependency, 95*, S5–S28. https://doi.org/10.1016/j.drugalcdep.2008.01.004

Kindermann, T. A. (2016). Peer group influences on students' academic motivation. In K. Wentzel & G. Ramani (Eds.), *Handbook of social influences in school contexts: social-emotional, motivational, and cognitive outcomes* (pp. 31–47). Routledge, Taylor & Francis.

Kindermann, T. A., & Skinner, E. A. (2012). Will the real peer group please stand up? A "tensegrity" approach to examining the synergistic influences of peer groups and friendship networks on academic development. In A. Ryan & G. Ladd (Eds.), *Peer relationships and adjustment at school* (pp. 51–78). Information Age Publishing.

Koomen, H. M. Y., & Hoeksma, J. B. (2003). Regulation of emotional security by children after entry to special and regular kindergarten classes. *Psychological Reports, 93*, 1319–1334. https://doi.org/10.2466/pr0.2003.93.3f.1319

Korpershoek, H., Harms, T., de Boer, H., van Kuijk, M., & Doolaard, S. (2016). A meta-analysis of the effects of classroom management strategies and classroom management programs on students' academic, behavioral, emotional, and motivational outcomes. *Review of Educational Research, 86*, 643–680. https://doi.org/10.3102/0034654315626799

Ladd, G. W. (2006). Peer rejection, aggressive or withdrawn behavior, and psychological maladjustment from ages 5 to 12: An examination of four predictive models. *Child Development, 77*(4), 822–846. https://doi.org/10.1111/j.1467-8624.2006.00905.x

Ladd, G. W., & Troop-Gordon, W. (2003). The role of chronic peer difficulties in the development of children's psychological adjustment problems. *Child Development, 74*, 1344–1367. https://doi.org/10.1111/1467-8624.00611

Ladd, G. W., Kochenderfer, B. J., & Coleman, C. C. (1996). Friendship quality as a predictor of young children's early school adjustment. *Child Development, 67*(3), 1103–1118. https://doi.org/10.2307/1131882

Lamarche, V., Brendgen, M., Boivin, M., Vitaro, F., Dionne, G., & Pérusse, D. (2007). Do friends' characteristics moderate the prospective links between peer victimization and reactive and proactive aggression? *Journal of Abnormal Child Psychology, 35*(4), 665–680. https://doi.org/10.1007/s10802-007-9122-7

Lansford, J. E., Malone, P. S., Dodge, K. A., Pettit, G. S., & Bates, J. E. (2010). Developmental cascades of peer rejection, social information processing biases, and aggression during middle childhood. *Development and Psychopathology, 22*(3), 593–602. https://doi.org/10.1017/S0954579410000301

Laursen, B., Bukowski, W. M., Aunola, K., & Nurmi, J.-E. (2007). Friendship moderates prospective associations between social isolation and adjustment problems in young children. *Child Development, 78*, 1395–1404. https://doi.org/10.1111/j.1467-8624.2007.01072.x

Leflot, G., van Lier, P., Onghena, P., & Colpin, H. (2010). The role of teacher behavior management in the development of disruptive behaviors: An intervention study with the Good Behavior Game. *Journal of Abnormal Child Psychology, 38*, 869–882. https://doi.org/10.1007/s10802-010-9411-4

Leflot, G., van Lier, P., Verschueren, K., Onghena, P., & Colpin, H. (2011). Transactional associations among teacher support, peer social preference, and child externalizing behavior: a four-wave longitudinal study. *Journal of Clinical Child and Adolescent Psychology, 40*(1), 87–99. https://doi.org/10.1080/15374416.2011.533409

Leflot, G., van Lier, P., Onghena, P., & Colpin, H. (2013). The role of children's on-task behavior in the prevention of aggressive behavior development and peer rejection: A randomized controlled study of the Good Behavior Game in Belgian elementary classrooms. *Journal of School Psychology, 51*, 187–199. https://doi.org/10.1016/j.jsp.2012.12.006

Lei, H., Cui, Y., & Chiu, M. M. (2016). Affective teacher-student relationships and students' externalizing behavior problems: A meta-analysis. *Frontiers in Psychology, 7*, 1–12. https://doi.org/10.3389/fpsyg.2016.01311

Ly, J., & Zhou, Q. (2016). Bidirectional associations between teacher-child relationship quality and Chinese American immigrant children's behavior problems. *Journal of Clinical Child & Adolescent Psychology, 47*(6), 954–966. https://doi.org/10.1080/15374416.2016.1183496

Ly, J., & Zhou, Q. (2018). Bidirectional associations between teacher–child relationship quality and Chinese American immigrant children's behavior problems. *Journal of Clinical Child & Adolescent Psychology, 47*, 954–966. https://doi.org/10.1080/15374416.2016.1183496

Maldonado-Carreño, C., & Votruba-Drzal, E. (2011). Teacher–child relationships and the development of academic and behavioral skills during elementary school: A within- and between-child analysis. *Child Development, 82*, 601–616. https://doi.org/10.1111/j.1467-8624.2010.01533.x

McGrath, K. F., & van Bergen, P. (2015). Who, when, why and to what end? Students at risk of negative student-teacher relationships and their outcomes. *Educational Research Review, 14*, 1–17. https://doi.org/10.1016/j.edurev.2014.12.001

McKinnon, D. P. (1994). Analysis of mediating variables in prevention and intervention research. In A. Cázares & L. A. Beatty (Eds.), *Scientific methods in prevention research* (NIDA research monograph 139. DHHS Pub. No. 94-3631) (pp. 127–153). U.S. Govt. Print. Office, Washington, DC.

McLaughlin, K. A., Sheridan, M. A., Tibu, F., Fox, N. A., Zeanah, C. H., & Nelson, C. A. (2015). Causal effects of the early caregiving environment on development of stress response systems in children. *Proceedings of the National Academy of Sciences, 112*(18), 5637–5642. https://doi.org/10.1073/pnas.1423363112

Mehus, C. J., Doty, J., Chan, G., Kelly, A. B., Hemphill, S., Toumbourou, J., & McMorris, B. J. (2018). Testing the Social Interaction Learning Model's applicability to adolescent substance misuse in an Australian context. *Substance Use and Misuse, 53*(11), 1859–1868. https://doi.org/10.1080/10826084.2018.1441307

Mercer, S. H., & DeRosier, M. E. (2008). Teacher preference, peer rejection, and student aggression: A prospective study of transactional influence and independent contributions to emotional adjustment and grades. *Journal of School Psychology, 46*(6), 661–685. https://doi.org/10.1016/j.jsp.2008.06.006

Michels, N., Sioen, I., Clays, E., De Buyzere, M., Ahrens, W., Huybrechts, I., … De Henauw, S. (2013). Children's heart rate variability as stress indicator: Association with reported stress and cortisol. *Biological Psychology, 94*(2), 433–440. https://doi.org/10.1016/j.biopsycho.2013.08.005

Monk, C., Kovelenko, P., Ellman, L. M., Sloan, R. P., Bagiella, E., Gorman, J. M., & Pine, D. S. (2001). Enhanced stress reactivity in paediatric anxiety disorders: Implications for future cardiovascular health. *International Journal of Neuropsychopharmacology, 4*(2), 199–206. https://doi.org/10.1017/S146114570100236X

Mulder, R. H., Walton, E., Neumann, A., Houtepen, L. C., Felix, J. F., Bakermans-Kranenburg, M. J., … Relton, C. L. (2020). Epigenomics of being bullied: changes in DNA methylation following bullying exposure. *Epigenetics, 15*(6-7), 750–764. https://doi.org/10.1080/15592294.2020.1719303

Nakamoto, J., & Schwartz, D. (2010). Is peer victimization associated with academic achievement? A meta-analytic review. *Social Development, 19*(2), 221–242. https://doi.org/10.1111/j.1467-9507.2009.00539.x

Nansel, T. R., Overpeck, M., Pilla, R. S., Ruan, W. J., Simons-Morton, B., & Scheidt, P. (2001). Bullying behaviors among US youth: Prevalence and association with psychosocial adjustment. *JAMA, 285*(16), 2094–2100. https://doi.org/10.1001/jama.285.16.2094

Nelson, R. M., & DeBacker, T. K. (2008). Achievement motivation in adolescents: The role of peer climate and best friends. *The Journal of Experimental Education, 76*(2), 170–189. https://doi.org/10.3200/JEXE.76.2.170-190

Newcomb, A. F., & Bagwell, C. L. (1995). Children's friendship relations: A meta-analytic review. *Psychological Bulletin, 117*, 306–347. https://doi.org/10.1037/0033-2909.117.2.306

O'Connor, E. E., & McCartney, K. (2007). Examining teacher–child relationships and achievement as part of an ecological model of development. *American Educational Research Journal, 44*, 340–369. https://doi.org/10.3102/0002831207302172

O'Connor, E. E., Collins, B. A., & Supplee, L. (2012). Behavior problems in late childhood: The roles of early maternal attachment and teacher-child relationship trajectories. *Attachment & Human Development, 14*, 265–288. https://doi.org/10.1080/14616734.2012.672280

Oldenburg, C. M., & Kerns, K. A. (1997). Associations between peer relationships and depressive symptoms: Testing moderator effects of gender and age. *Journal of Early Adolescence, 17*, 319–337. https://doi.org/10.1177/0272431697017003004

Olweus, D. (1978). *Aggression in the schools: Bullies and whipping boys*. Hemisphere.

Ouellet-Morin, I., Danese, A., Bowes, L., Shakoor, S., Ambler, A., Pariante, C., Papadopoulos, A., Caspi, A., Moffitt, T. E., & Arseneault, L. (2011). A discordant monozygotic twin design shows blunted cortisol reactivity among bullied children. *Journal of the American Academy of Child and Adolescent Psychiatry, 50*(6), 574–582 e3. https://doi.org/10.1016/j.jaac.2011.02.015

Ouellet-Morin, I., Wong, C., Danese, A., Pariante, C., Papadopoulos, A., Mill, J., & Arseneault, L. (2013). Increased serotonin transporter gene (SERT) DNA methylation is associated with bullying victimization and blunted cortisol response to stress in childhood: a longitudinal study of discordant monozygotic twins. *Psychological Medicine, 43*(9), 1813. https://doi.org/10.1017/S0033291712002784

Pahigiannis, K., & Glos, M. (2020). Peer influences in self-regulation development and interventions in early childhood. *Early Child Development and Care, 190*, 1053–1064. https://doi.org/10.1080/03004430.2018.1513923

Pakarinen, E., Aunola, K., Kiuru, N., Lerkkanen, M.-K., Poikkeus, A.-M., Siekkinen, M., & Nurmi, J.-E. (2014). The cross-lagged associations between classroom interactions and children's achievement behaviors. *Contemporary Educational Psychology, 39*, 248–261. https://doi.org/10.1016/j.cedpsych.2014.06.001

Pakarinen, E., Silinskas, G., Hamre, B. K., Metsäpelto, R.-L., Lerkkanen, M.-K., Poikkeus, A.-M., & Nurmi, J.-E. (2018). Cross-lagged associations between problem behaviors and teacher-student relationships in early adolescence. *Journal of Early Adolescence, 38*, 1100–1141. https://doi.org/10.1177/0272431617714328

Parker, J. G., & Asher, S. R. (1993). Friendship and friendship quality in middle childhood: Links with peer group acceptance and feelings of loneliness and social dissatisfaction. *Developmental Psychology, 29*, 611–621. https://doi.org/10.1037/0012-1649.29.4.611

Parker, J. G., & Seal, J. (1996). Forming, losing, renewing, and replacing friendships: Applying temporal parameters to the assessment of children's friendship experiences. *Child Development, 67*, 2248–2268. https://doi.org/10.2307/1131621

Parker, J. G., Rubin, K. H., Erath, S. A., Wojslawowicz, J. C., & Buskirk, A. A. (2015). Peer relationships, child development, and adjustment: A developmental psychopathology perspective. In D. Cicchetti & D. J. Cohen (Eds.), *Developmental psychopathology*. Wiley.

Pedersen, S., Vitaro, F., Barker, E. D., & Borge, A. I. H. (2007). The timing of middle-childhood peer rejection and friendship: Linking early behavior to early-adolescent adjustment. *Child Development, 78*(4), 1037–1051. https://doi.org/10.1111/j.1467-8624.2007.01051.x

Peters, E., Riksen-Walraven, J. M., Cillessen, A. H. N., & de Weerth, C. (2011). Peer rejection and HPA activity in middle childhood: Friendship makes a difference. *Child Development, 82*(6), 1906–1920. https://doi.org/10.1111/j.1467-8624.2011.01647.x

Pianta, R. C. (2001). *STRS: Student-teacher relationship scale: Professional manual*. Psychological Assessment Resources.

Pianta, R. C., Hamre, B. K., & Stuhlman, M. (2003). Relationships between teachers and children. In W. M. Reynolds, G. E. Miller, & I. B. Weiner (Eds.), *Handbook of psychology* (Educational psychology) (Vol. 7, pp. 199–234). John Wiley.

Pianta, R. C., & Stuhlman, M. W. (2004). Teacher-child relationships and children's success in the first years of school. *School Psychology Review, 33*, 444–458. https://doi.org/10.1002/0471264385.wei0710

Pianta, R. C., Belsky, J., Vandergrift, N., Houts, R., & Morrison, F. J. (2008). Classroom effects on children's achievement trajectories in elementary school. *American Educational Research Journal, 45*, 365–397. https://doi.org/10.3102/0002831207308230

Piehler, T. F., & Dishion, T. J. (2007). Interpersonal dynamics within adolescent friendships: Dyadic mutuality, deviant talk, and patterns of antisocial behavior. *Child Development, 78*(5), 1611–1624. https://doi.org/10.1111/j.1467-8624.2007.01086.x

Poulin, F., Dishion, T. J., & Haas, E. (1999). The peer influence paradox: Friendship quality and deviancy training within male adolescent friendships. *Merrill-Palmer Quarterly, 45*, 42–61.

Pouwels, J. L., Souren, P. M., Lansu, T. A., & Cillessen, A. H. (2016). Stability of peer victimization: A meta-analysis of longitudinal research. *Developmental Review, 40*, 1–24. https://doi.org/10.1016/j.dr.2016.01.001

Prinstein, M. J. (2007). Moderators of peer contagion: A longitudinal examination of depression socialization between adolescents and their best friends. *Journal of Clinical Child and Adolescent Psychology, 36*(2), 159–170. https://doi.org/10.1080/15374410701274934

Reijntjes, A., Kamphuis, J. H., Prinzie, P., & Telch, M. J. (2010). Peer victimization and internalizing problems in children: A meta-analysis of longitudinal studies. *Child Abuse & Neglect, 34*(4), 244–252. https://doi.org/10.1016/j.chiabu.2009.07.009

Reijntjes, A., Kamphuis, J. H., Prinzie, P., Boelen, P. A., Van der Schoot, M., & Telch, M. J. (2011). Prospective linkages between peer victimization and externalizing problems in children: A meta-analysis. *Aggressive Behavior, 37*(3), 215–222. https://doi.org/10.1002/ab.20374

Roorda, D. L., Koomen, H. M. Y., Spilt, J. L., & Oort, F. J. (2011). The influence of affective teacher-student relationships on students' school engagement and achievement: A meta-analytic approach. *Review of Educational Research, 81*, 493–529. https://doi.org/10.3102/0034654311421793

Roorda, D. L., Verschueren, K., Vancraeyveldt, C., Van Craeyevelt, S., & Colpin, H. (2014). Teacher-child relationships and behavioral adjustment: Transactional links for preschool boys at risk. *Journal of School Psychology, 52*, 495–510. https://doi.org/10.1016/j.jsp.2014.06.004

Roorda, D. L., Jak, S., Zee, M., Oort, F. J., & Koomen, H. M. Y. (2017). Affective teacher-student relationships and students' engagement and achievement: A meta-analytic update and test of the mediating role of engagement. *School Psychology Review, 46*, 239–261. https://doi.org/10.17105/SPR-2017-0035.V46-3

Roorda, D. L., Zee, M., & Koomen, H. M. Y. (2021). Don't forget student-teacher dependency! A meta-analysis on associations with students' school adjustment and the moderating role of student and teacher characteristics. *Attachment & Human Development, 23* (5), 490-503. https://doi.org/10.1080/14616734.2020.1751987

Rose, A. J. (2002). Co-rumination in the friendship of girls and boys. *Child Development, 73*, 1830–1843. https://doi.org/10.1111/1467-8624.00509

Rubin, K. H., Bukowski, W. M., & Parker, J. G. (2006). Peer interactions, relationships, and groups. In N. Eisenberg (Ed.), *Handbook of child psychology. Vol 3 social, emotional, and personality development* (pp. 571–645). Wiley.

Rucinski, C. L., Brown, J. L., & Downer, J. T. (2018). Teacher-child relationships, classroom climate, and children's social-emotional and academic development. *Journal of Educational Psychology, 110*, 992–1104. https://doi.org/10.1037/edu0000240

Rudolph, K. D., Troop-Gordon, W., & Flynn, M. (2009). Relational victimization predicts children's social-cognitive and self-regulatory responses in a challenging peer context. *Developmental Psychology, 45*(5), 1444–1454. https://doi.org/10.1037/a0014858

Rueger, S. Y., Malecki, C. K., Pyun, Y., Aycock, C., & Coyle, S. (2016). A meta-analytic review of the association between perceived social support and depression in childhood and adolescence. *Psychological Bulletin, 142*, 1017–1067. https://doi.org/10.1037/bul0000058

Saarento, S., Boulton, A. J., & Salmivalli, C. (2015). Reducing bullying and victimization: Student- and classroom-level mechanisms of change. *Journal of Abnormal Child Psychology, 43*, 61–76. https://doi.org/10.1007/s10802-013-9841-x

Sabol, T. J., & Pianta, R. C. (2012). Recent trends in research on teacher–child relationships. *Attachment & Human Development, 14*, 213–231. https://doi.org/10.1080/14616734.2012.672262

Salmivalli, C., & Isaacs, J. (2005). Prospective relations among victimization, rejection, friendlessness, and children's self- and peer-perceptions. *Child Development, 76*(6), 1161–1171. https://doi.org/10.1111/j.1467-8624.2005.00841.x-i1

Salvas, M.-C., Vitaro, F., Brendgen, M., Lacourse, E., Boivin, M., & Tremblay, R. E. (2011). Interplay between friends' aggression and friendship quality in the development of child aggression during the early school years. *Social Development, 20*(4), 645–663. https://doi.org/10.1111/j.1467-9507.2010.00592.x

Salvas, M.-C., Vitaro, F., Brendgen, M., Dionne, G., Tremblay, R. E., & Boivin, M. (2014). Friendship conflict and the development of generalized physical aggression in the early school years: A genetically informed study of potential moderators. *Developmental Psychology, 50*(6), 1794–1807. https://doi.org/10.1037/a0036419

Salvas, M.-C., Vitaro, F., Brendgen, M., & Cantin, S. (2016). Prospective links between friendship and early physical aggression: Preliminary evidence supporting the role of friendship quality through a dyadic intervention. *Merrill-Palmer Quarterly, 62*(3), 285–305.

Schwartz, D. (2000). Subtypes of victims and aggressors in children's peer groups. *Journal of Abnormal Child Psychology, 28*(2), 181–192. https://doi.org/10.1023/A:1005174831561

Schwartz, D., Gorman, A. H., Nakamoto, J., & Toblin, R. L. (2005). Victimization in the peer group and children's academic functioning. *Journal of Educational Psychology, 97*(3), 425. https://doi.org/10.1037/0022-0663.97.3.425

Sebanc, A. M., Guimond, A. B., & Lutgen, J. (2016). Transactional relationships between Latinos' friendship quality and academic achievement during the transition to middle school. *Journal of Early Adolescence, 36*, 108–138. https://doi.org/10.1177/0272431614556347

Serdiouk, M., Rodkin, P., Madill, R., Logis, H., & Gest, S. (2015). Rejection and victimization among elementary school children: The buffering role of classroom-level predictors. *Journal of Abnormal Child Psychology, 43*, 5–17. https://doi.org/10.1007/s10802-013-9826-9

Sette, S., Gasser, L., & Grütter, J. (2020). Links between teachers' liking of students, peer inclusion, and students' academic achievement: A two-wave longitudinal study. *Journal of Youth and Adolescence, 43*, 747–756. https://doi.org/10.1007/s10964-019-01048-5

Shin, H., & Ryan, A. M. (2017). Friend influence on early adolescent disruptive behavior in the classroom: Teacher emotional support matters. *Developmental Psychology, 53*, 114–125. https://doi.org/10.1037/dev0000250

Skalická, V., Stenseng, F., & Wichstrøm, L. (2015). Reciprocal relations between student–teacher conflict, children's social skills and externalizing behavior: A three-wave longitudinal study

from preschool to third grade. *International Journal of Behavioral Development, 39*, 413–425. https://doi.org/10.1177/0165025415584187
Snyder, J., Schrepferman, L., Oeser, J., Patterson, G., Stoolmiller, M., Johnson, K., & Snyder, A. (2005). Deviancy training and association with deviant peers in young children: Occurrence and contribution to early-onset conduct problems. *Development and Psychopathology, 17*(2), 397–413. https://doi.org/10.1017/S0954579405050194
Snyder, J., Schrepferman, L., Stoolmiller, M., & Brooker, M. (2007). The roles of anger, conflict with parents and peers, and social reinforcement in the early development of physical aggression. In T. A. Cavell & K. T. Malcolm (Eds.), *Anger, aggression, and interventions for interpersonal violence* (pp. 187–214). Erlbaum.
Spilt, J. L., Hughes, J. N., Wu, J. Y., & Kwok, O. M. (2012a). Dynamics of teacher–student relationships: Stability and change across elementary school and the influence on children's academic success. *Child Development, 83*, 1180–1195. https://doi.org/10.1111/j.1467-8624.2012.01761.x
Spilt, J. L., Koomen, H. M. Y., & Jak, S. (2012b). Are boys better off with male and girls with female teachers? A multilevel investigation of measurement invariance and gender match in teacher-student relationship quality. *Journal of School Psychology, 50*, 363–378. https://doi.org/10.1016/j.jsp.2011.12.002
Spilt JL, van Lier PA, Leflot G, Onghena P, Colpin H. (2014). Children's social self-concept and internalizing problems: The influence of peers and teachers. *Child Development, 85*(3), 1248–1256. https://doi.org/10.1111/cdev.12181
Spilt, J. L., Leflot, G., Onghena, P., & Colpin, H. (2016). Use of praise and reprimands as critical ingredients of teacher behavior management: Effects on children's development in the context of a teacher-mediated classroom intervention. *Prevention Science, 17*, 732–742. https://doi.org/10.1007/s11121-016-0667-y
Spilt, J. L., Leflot, G., & Colpin, H. (2019). Teacher involvement prevents increases in depressive symptoms: Bidirectional associations in elementary school. *Journal of Abnormal Child Psychology, 47*, 359–367. https://doi.org/10.1007/s10802-018-0441-7
Stenseng, F., Belsky, J., Skalicka, V. & Wichstrom, L. (2015). Social exclusion predicts impaired self-regulation: A 2-year longitudinal panel study including transition from preschool to school. *Journal of Personality, 83*(2), 212–220. https://doi.org/10.1111/jopy.12096
Stevens, E. A., & Prinstein, M. J. (2005). Peer contagion of depressogenic attributional styles among adolescents: A longitudinal study. *Journal of Abnormal Child Psychology, 33*(1), 25–37. https://doi.org/10.1007/s10802-005-0931-2
Sturaro, C., van Lier, P. A. C., Cuijpers, P., & Koot, H. M. (2011). The role of peer relationships in the development of early school-age externalizing problems. *Child Development, 82*(3), 758–765. https://doi.org/10.1111/j.1467-8624.2010.01532.x
Sullivan, H. S. (1953). *The interpersonal theory of psychiatry*. Norton.
Takizawa, R., Maughan, B., & Arseneault, L. (2014). Adult health outcomes of childhood bullying victimization: Evidence from a five-decade longitudinal British birth cohort. *American Journal of Psychiatry, 171*, 777–784. https://doi.org/10.1176/appi.ajp.2014.13101401
Taylor, A. R. (1989). Predictors of peer rejection in early elementary grades: Roles of problem behavior, academic achievement, and teacher preference. *Journal of Clinical Child Psychology, 18*(4), 360–365. https://doi.org/10.1207/s15374424jccp1804_10
Teräsahjo, T., & Salmivalli, C. (2003). "She is not actually bullied." The discourse of harassment in student groups. *Aggressive Behavior, 29*(2), 134–154. https://doi.org/10.1002/ab.10045
Thompson, E. L., Mehari, K. R., & Farrell, A. D. (2019). Deviant peer factors during early adolescence: Cause or consequence of physical aggression? *Child Development, 91*. https://doi.org/10.1111/cdev.13242
Tieskens, J. M., Buil, J. M., Koot, S., & van Lier, P. A. (2019). Relational victimization and elementary schoolchildren's risk-taking behavior: Impact of the classroom norm toward risk-taking. *International Journal of Behavioral Development*. https://doi.org/10.1177/0165025419880617

Trentacosta, C, & Shaw, D. (2009). Emotional self-regulation, peer rejection, and antisocial behavior from early childhood to early adolescence. *Journal of Applied Developmental Psychology, 30*(3), 356–365. https://doi.org/10.1016/j.appdev.2008.12.016

Troop-Gordon, W., & Kopp, J. (2011). Teacher-child relationship and children's peer victimization and aggressive behavior in late childhood. *Social Development, 20*, 536–561. https://doi.org/10.1111/j.1467-9507.2011.00604.x

Vaillancourt, T., Brittain, H. L., McDougall, P., & Duku, E. (2013a). Longitudinal links between childhood peer victimization, internalizing and externalizing problems, and academic functioning: Developmental cascades. *Journal of Abnormal Child Psychology, 41*(8), 1203–1215. https://doi.org/10.1007/s10802-013-9781-5

Vaillancourt, T., Hymel, S., & McDougall, P. (2013b). The biological underpinnings of peer victimization: Understanding why and how the effects of bullying can last a lifetime. *Theory into Practice, 52*(4), 241–248. https://doi.org/10.1080/00405841.2013.829726

Van Craeyevelt, S., Verschueren, K., Vancraeyveldt, C., Wouters, S., & Colpin, H. (2017). The role of preschool teacher–child interactions in academic adjustment: An intervention study with Playing-2-gether. *British Journal of Educational Psychology, 87*, 345–364. https://doi.org/10.1111/bjep.12153

van der Meulen, M., Steinbeis, N., Achterberg, M., Bilo, E., van den Bulk, B. G., & van IJzendoorn, M. H., & Crone, E. A. (2017). The neural correlates of dealing with social exclusion in childhood. *Neuropsychologia, 103*, 29–37. https://doi.org/10.1016/j.neuropsychologia.2017.07.008

van der Meulen, M., Steinbeis, N., Achterberg, M., van Ijzendoorn, M. H., & Crone, E. A. (2018). Heritability of neural reactions to social exclusion and prosocial compensation in middle childhood. *Developmental Cognitive Neuroscience, 34*, 42–52. https://doi.org/10.1016/j.dcn.2018.05.010

van Lier, P. A., & Koot, H. M. (2010). Developmental cascades of peer relations and symptoms of externalizing and internalizing problems from kindergarten to fourth-grade elementary school. *Development and Psychopathology, 22*(3), 569–582. https://doi.org/10.1017/S0954579410000283

Van Lier, P. A. C., Vuijk, P., & Crijnen, A. A. M. (2005). Understanding mechanisms of change in the development of antisocial behavior: Impact of a universal intervention. *Journal of Abnormal Child Psychology, 33*, 521–535. https://doi.org/10.1007/s10802-005-6735-7

van Lier, P., Boivin, M., Dionne, G., Vitaro, F., Brendgen, M., Koot, H., ... Pérusse, D. (2007). Kindergarten children's genetic vulnerabilities interact with friends' aggression to promote children's own aggression. *Journal of the American Academy of Child and Adolescent Psychiatry, 46*(8), 1080–1087. https://doi.org/10.1097/CHI.0b013e318067733e

van Lier, P. A. C., Vitaro, F., Barker, E. D., Brendgen, M., Tremblay, R. E., & Boivin, M. (2012). Peer victimization, poor academic achievement, and the link between childhood externalizing and internalizing problems. *Child development, 83*(5), 1775–1788. https://doi.org/10.1111/j.1467-8624.2012.01802.x

Vancraeyveldt, C., Verschueren, K., Wouters, S., Van Craeyevelt, S., Van Den Noortgate, W., & Colpin, H. (2015). Improving teacher-child relationship quality and teacher-rated behavioral adjustment amongst externalizing preschoolers: Effects of a two-component intervention. *Journal of Abnormal Child Psychology, 43*, 243–257. https://doi.org/10.1007/s10802-014-9892-7

Vandenbroucke, L., Spilt, J. L., Verschueren, K., & Baeyens, D. (2018). The effects of peer rejection, parent and teacher support on working memory performance: An experimental approach in middle childhood. *Learning and Individual Differences, 67*, 12–21. https://doi.org/10.1016/j.lindif.2018.06.007

Veenman, B., Luman, M., & Oosterlaan, J. (2018). Efficacy of behavioral classroom programs in primary school. A meta-analysis focusing on randomized controlled trials. *PLOS One, 13*(10), e0201779. https://doi.org/10.1371/journal.pone.0201779

Veenstra, R., Dijkstra, J. K., & Kreager, D. A. (2018). Pathways, networks, and norms: A sociological perspective on peer research. In W. M. Bukowski, B. Laursen, & K. H. Rubin (Eds.), *Handbook of peer interactions, relationships, and groups* (pp. 45–63). The Guilford Press.

Vernon-Feagans, L., Mokrova, I. L., Carr, R. C., Garrett-Peters, P. T., Burchinal, M. R., & The Family Life Project Key Investigators. (2019). Cumulative years of classroom quality from kindergarten to third grade: Prediction to children's third grade literacy skills. *Early Childhood Research Quarterly, 47*, 531–540. https://doi.org/10.1016/j.ecresq.2018.06.005

Véronneau, M.-H., Vitaro, F., Pedersen, S., & Tremblay, R. E. (2008). Do peers contribute to the likelihood of secondary school graduation among disadvantaged boys? *Journal of Educational Psychology, 100*(2), 429–442. https://doi.org/10.1037/0022-0663.100.2.429

Véronneau, M.-H., Vitaro, F., Brendgen, M., Dishion, T. J., & Tremblay, R. E. (2010). Transactional analysis of the reciprocal links between peer experiences and academic achievement from middle childhood to early adolescence. *Developmental Psychology, 46*(4), 773–790. https://doi.org/10.1037/a0019816

Verschueren, K. (2015). Middle childhood teacher-child relationships: Insights from an attachment perspective and remaining challenges. In G. Bosmans & K.A. Kerns (Eds.), Attachment in middle childhood: Theoretical challenges and new directions in an emerging field. *New Directions for Child and Adolescent Development, 148*, 77–91. https://doi.org/10.1002/cad.20097

Verschueren, K., & Koomen, H. M. Y. (2012). Teacher-child relationships from an attachment perspective. *Attachment and Human Development, 14*, 205–211. https://doi.org/10.1080/14616734.2012.672260

Vitaro, F., Tremblay, R. E., Kerr, M., Pagani, L. S., & Bukowski, W. M. (1997). Disruptiveness, friends' characteristics, and delinquency: A test of two competing models of development. *Child Development, 68*(4), 676–689. https://doi.org/10.2307/1132118

Vitaro, F., Pedersen, S., & Brendgen, M. (2007). Children's disruptiveness, peer rejection, friends' deviancy, and delinquent behaviors: A process-oriented approach. *Development and Psychopathology, 19*(2), 433–453. https://doi.org/10.1017/S0954579407070216

Vitaro, F., Boivin, M., & Bukowski, W. M. (2009a). The role of friendship in child and adolescent psychosocial development. In K. Rubin, W. M. Bukowski, & B. Laursen (Eds.), *Handbook of peer interactions, relationships, and groups* (pp. 568–588). Guilford Press.

Vitaro, F., Brendgen, M., & Arseneault, L. (2009b). The discordant MZ-twin method: One step closer to the holy grail of causality. *International Journal of Behavioral Development, 33*(4), 376–382. https://doi.org/10.1177/0165025409340805

Vitaro, F., Brendgen, M., Boivin, M., Cantin, S., Dionne, G., Tremblay, R. E., … Pérusse, D. (2011). A monozygotic twin difference study of friends' aggression and children's adjustment problems. *Child Development, 82*(2), 617–632. https://doi.org/10.1111/j.1467-8624.2010.01570.x

Vitaro, F., Boivin, M., Brendgen, M., Girard, A., & Dionne, G. (2012). Social experiences in kindergarten and academic achievement in grade 1: A Monozygotic Twin Difference Study. *Journal of Educational Psychology, 104*(2), 366–380. https://doi.org/10.1037/a0026879

Vitaro, F., Brendgen, M., Girard, A., Boivin, M., Dionne, G., & Tremblay, R. E. (2015). The expression of genetic risk for aggressive and non-aggressive antisocial behavior is moderated by peer group norms. *Journal of Youth and Adolescence, 44*(7), 1379–1395. https://doi.org/10.1007/s10964-015-0296-y

Vitaro, F., Brendgen, M., Girard, A., Dionne, G., Tremblay, R. E., & Boivin, M. (2016). Links between friends' physical aggression and adolescents' physical aggression: What happens if gene-environment correlations are controlled? *International Journal of Behavioral Development, 40*(3), 234–242. https://doi.org/10.1177/0165025415618277

Vitaro, F., Boivin, M., & Poulin, F. (2018). The interface of aggression and peer relations in childhood and adolescence. In W. M. Bukowski, B. Laursen, & K. Rubin (Eds.), *Handbook of peer interactions, relationships, and groups* (2nd ed., pp. 284–301). Guilford Press.

Wentzel, K. R. (1996). Social goals and social relationships as motivators of school adjustment. In J. Juvonen & K. R. Wentzel (Eds.), *Social motivation: Understanding children's school adjustment* (pp. 226–247). Cambridge University Press. https://doi.org/10.1017/CBO9780511571190.012

Wentzel, K. R. (2009). Peers and academic functioning at school. In K. H. Rubin, W. M. Bukowski, & B. Laursen (Eds.), *Handbook of peer interactions, relationships, and groups* (pp. 531–547). Guilford Press.

Wentzel, K. R., & Muenks, K. (2016). Peer influence on students' motivation, academic achievement, and social behavior. In K. Wentzel & G. Ramani (Eds.), *Handbook of social influences in school contexts: Social-emotional, motivation and cognitive outcomes* (pp. 13–30). Routledge.

Wentzel, K. R., Barry, C. M., & Caldwell, K. A. (2004). Friendships in middle school: Influences on motivation and school adjustment. *Journal of Educational Psychology, 96*(2), 195–203. https://doi.org/10.1037/0022-0663.96.2.195

Weyns, T., Verschueren, K., Leflot, G., Onghena, P., Wouters, S., & Colpin, H. (2017). The role of teacher behavior in children's relational aggression development: A five-wave longitudinal study. *Journal of School Psychology, 64*, 17–27. https://doi.org/10.1016/j.jsp.2017.04.008

Weyns, T., Colpin, H., De Laet, S., Engels, M., & Verschueren, K. (2018). Teacher support, peer acceptance, and engagement in the classroom: A three-wave longitudinal study in late childhood. *Journal of Youth and Adolescence, 47*(6), 1139–1150. https://doi.org/10.1007/s10964-017-0774-5

Weyns, T., Colpin, H., Engels, M. C., Doumen, S., & Verschueren, K. (2019). The relative contribution of peer acceptance and individual and class-level teacher–child interactions to kindergartners' behavioral development. *Early Childhood Research Quarterly, 47*, 259–270. https://doi.org/10.1016/j.ecresq.2018.12.009

White, K. J., & Jones, K. (2000). Effects of teacher feedback on the reputations and peer perceptions of children with behavior problems. *Journal of Experimental Child Psychology, 76*(4), 302–326. https://doi.org/10.1006/jecp.1999.2552

White, K. J., Sherman, M. D., & Jones, K. (1996). Children's perceptions of behavior problem peers: Effects of teacher feedback and peer-reputed status. *Journal of School Psychology, 34*(1), 53–72. https://doi.org/10.1016/0022-4405(95)00025-9

Williams, K. D. (2007). Ostracism. *Annual Review of Psychology, 58*, 425–452. https://doi.org/10.1146/annurev.psych.58.110405.085641

Williford, A. P., LoCasale-Crouch, J., Whittaker, J. V., DeCoster, J., Hartz, K. A., Carter, L. M., Wolcott, C. S., & Hatfield, B. E. (2017). Changing teacher-child dyadic interactions to improve preschool children's externalizing behaviors. *Child Development, 88*, 1544–1553. https://doi.org/10.1111/cdev.12703

Wolke, D., Woods, S., Bloomfield, L., & Karstadt, L. (2001). Bullying involvement in primary school and common health problems. *Archives of Disease in childhood, 85*(3), 197–201. https://doi.org/10.1136/acd.85.3.197

Zee, M., Koomen, H. M. Y., & van der Veen, I. (2013). Student-teacher relationship quality and academic adjustment in upper elementary school: The role of child personality. *Journal of School Psychology, 51*, 517–533. https://doi.org/10.1016/j.jsp.2013.05.003

School Social Relations and Child Development: Gene-Environment Interplay

Mara Brendgen and Tina Kretschmer

Introduction

For a large part of the past century, most theoretical perspectives on human development have focused on the environment as a source of individual differences. There is now widespread consensus, however, that genetic and environmental factors shape all aspects of human development (e.g., Plomin et al., 2016). These forces interact and reciprocally influence each other through various mechanisms of gene-environment interplay (Rutter et al., 2006). Understanding this interplay is crucial for a thorough knowledge of child development. To date, the vast majority of studies have examined how genetic factors work together with family-related factors (e.g., socioeconomic status, parenting) or stressful life events and experiences (e.g., maltreatment) to explain differences in developmental outcomes. Already at a fairly young age, however, children spend many hours in educational settings, which provide opportunities to interact with other important socializing agents – notably teachers and peers – who can influence their developmental adjustment and well-being. Especially in primary school, teachers not only transmit academic and socio-cultural knowledge but also play a critical role in facilitating and managing the social interactions among the children in the class (Kindermann, 2011). For their part, peers are believed to provide a unique context for the acquisition of social skills, the validation of the self-concept, and the learning of social roles, norms, and behaviors (Boivin et al., 2005).

M. Brendgen
Département de Psychologie, Université du Québec à Montréal, Montreal, QC, Canada
e-mail: Brendgen.Mara@uqam.ca

T. Kretschmer (✉)
Department of Pedagogy and Educational Sciences, University of Groningen, Groningen, The Netherlands
e-mail: t.kretschmer@rug.nl

P. A. C. van Lier, K. Deater-Deckard (eds.), *Biosocial Interplay During Elementary School*, https://doi.org/10.1007/978-3-031-07109-6_4

Recently, researchers have started documenting how genetic factors work together with the social school environment to influence child development. This chapter reviews the current state of knowledge in this context and covers research on primary school-aged children, documenting both relationship-based (i.e., teacher-child relationship, school friendships, peer acceptance and rejection, victimization) and context-based (e.g., school policies, classroom norms) factors of the school environment, as well as two main maladjustment outcomes that so far have received the most attention (i.e., externalizing and internalizing symptoms). We first describe the various mechanisms of gene-environment interplay as they may relate to links between the social school environment and child adjustment. Next, we provide a brief illustration of quantitative genetic and molecular genetic approaches, followed by an overview of evidence of genetic influence on externalizing and internalizing problems. Recent findings that illustrate gene-and-school environment interplay in regard to the maladaptive outcomes of interest are presented subsequently. We focus specifically on empirical studies conducted with children up to approximately age 12, as this constitutes the average age at which children transition to middle school. We conclude the chapter by discussing the implications for theory and practice and by outlining future directions in genetically informed research on the association between social relations in school and child development.

Gene-Environment Interplay

The term "gene-environment interplay" can refer to a variety of concepts (Rutter et al., 2006). However, the two mechanisms most relevant for understanding the link between the school social environment and children's behavioral, emotional, and academic development are *gene-environment correlations* and *gene-environment interactions*.

Gene-Environment Correlations (rGE)

Gene-environment correlations (denoted rGE) refer to a situation where heritable factors are associated with specific environments. Of particular theoretical and practical interest in the context of this chapter is the type of rGE that arises when a child's genetically influenced traits (e.g., behaviors, cognitive abilities) affect the environment he or she will experience. Scarr and McCartney (1983) have described three ways how genetic factors can influence individuals' environmental experiences. *Passive* rGE occurs when parents' personal characteristics, which are partly explained by genetic factors, influence the environment they provide for their children. For example, parents with a strong athletic aptitude (a highly heritable trait, Costa et al., 2012) may be more likely to enroll their child in a sports enrichment program in school, thereby also influencing the kind of teachers and peers the child is exposed to. The child's genotype, which is inherited from the parents, becomes correlated with his or her social environment in school. *Passive rGE* may thus underlie parental influence on children's relations with teachers and peers.

In contrast to passive rGE, which involves the parents' selecting or structuring of the child's environment, *selective rGE* and *evocative rGE* involve environmental features that are presumably influenced by the child's heritable characteristics. *Selective* (also called *active*) rGE arises when individuals actively select or shape their own environments based on their genetically influenced personal characteristics. Selective rGE occurs, for example, when aggressive youths (whose behavior is in part genetically influenced) actively select friends with similar behavioral characteristics.

Finally, *evocative* (also called *re-active*) rGE occurs when the child's genetically influenced characteristics lead to specific reactions from the social environment. For example, a genetic disposition for aggression may elicit negative reactions from teachers and rejection from classmates, thus generating a correlation between a genetic risk for aggression and children's social relations with teachers and peers. Importantly, the presence of rGE does not imply that problematic social relations in school cannot also cause aggression but rather that the association between both is partly accounted for by genetic factors.

Gene-Environment Interactions (GxE)

GxE refers to a process whereby (a) the expression of a genetic disposition toward a developmental outcome varies as a function of the environment or (b) the effect of the environment varies depends on an individual's genetic disposition (Shanahan & Hofer, 2005). Like rGE, GxE may arise through different processes. A *trigger process* occurs when an environmental condition triggers or exacerbates a genetic predisposition for a given outcome or when it leads to a specific outcome only in individuals with the predisposing genes. When involving "adverse" environmental conditions and negative outcomes, such as when victimization by peers in school leads to aggression mainly (or only) in individuals with a genetic risk for such behavior, this *trigger process* is an analog to the diathesis-stress model (Rende & Plomin, 1992). However, a *trigger process* of GxE may also involve an exceptionally supportive environment and positive outcomes, such as when enrolment in an educational enrichment program allows an academically gifted child to reach his or her full potential. This type of GxE has been referred to as an enhancement process (Shanahan & Hofer, 2005) or as vantage sensitivity (Pluess & Belsky, 2013).

Both trigger processes are conceived as involving specific genetic susceptibility factors. However, the same genetic disposition may increase sensitivity to both very negative and very positive environments, such that a child who is especially vulnerable to adverse conditions may also benefit most from positive environments (Belsky & Pluess, 2009). This type of U-shaped trigger GxE effect, which is referred to as *differential susceptibility*, may occur in children who are genetically predisposed to showing greater physiological reactivity to both reward and punishment.

A *suppression process* of GxE may arise when environmental conditions reduce the role of genetic factors. As in the case of a trigger process, a suppression process of GxE can involve "adverse" environmental conditions, such as exposure to war or famine, which may trigger aggression or depression in a large number of individuals regardless of their genetic disposition. Again, however, a suppression process can also involve environmental conditions that might be considered positive, such as social norms or constraints that limit the expression of individuals' genetic vulnerability for problem behavior. Such a situation has also been called a *compensation or social control process* of GxE (Shanahan & Hofer, 2005).

Methods of Testing Gene-Environment Interplay

While most empirical evidence for gene-environment interplay regarding the school social environment comes from quantitative genetic studies, molecular genetic studies are on the rise. Both allow testing rGE and GxE. However, whereas quantitative genetic studies statistically infer genetic effects without collecting DNA, molecular genetic studies collect biological samples (e.g., from blood or saliva) to identify particular genes related to a behavior or symptom.

Quantitative Genetic Studies

In quantitative genetic studies, the relative strength of genetic and environmental effects is statistically deduced by examining the phenotypic similarity of family members who vary in their genetic relatedness. Quantitative studies use different research designs, such as the comparison of adopted and biological siblings or the comparison of identical (monozygotic) and fraternal (dizygotic) twin pairs growing up together. Common to all of these designs is the assumption that inter-individual differences in a measured outcome (i.e., a phenotype) can be decomposed into three sources of variance: **genetic (or heritable) factors**, **shared (or common) environmental factors**, and **nonshared (or unique) environmental factors** (Røysamb & Tambs, 2016).

Genes can come in different forms or variants (called alleles), which are located at the same position, or genetic locus, on a chromosome. In an adoption design, genetic influences are thought to explain interindividual differences if individuals are "phenotypically" more similar to their full biological siblings – with whom they share, by descent, on average 50% of allelic variants – than to their genetically unrelated adopted or stepsiblings. In the classical twin design, genetic factors are assumed to be implicated if identical or monozygotic (MZ) twins, who by descent share 100% of their allelic variants (barring mutations), are more similar to each other than nonidentical or dizygotic (DZ) twins, who by descent share on average only 50% of allelic variants, just like full siblings. Shared environment encompasses environmental influences (including those occurring prenatally) that affect siblings in the same way, thus making them similar to each other. Importantly, shared environment can refer to features both within (e.g., family revenue, parental education) and outside the family (e.g., neighborhood characteristics). When siblings attend the same classroom, the shared environment may also include the school social environment. Indeed, although rates vary considerably across countries, studies suggest that at least a third of twin pairs attend the same classroom during elementary school (Kovas et al., 2007; Polderman et al., 2010; White et al., 2018). Finally, nonshared environment refers to any experiences within or outside the family (or environmental experiences occurring prenatally) that make siblings dissimilar. For example, parents may treat siblings differently, which in turn may result in behavioral differences (Conger & Conger, 1994). The most important nonshared environmental influences, however, are likely those experienced outside the family (Dunn & Plomin, 1990; Plomin, 2011). For the majority of twins who are in separate classrooms, this includes their relations with teachers and peers. Moreover, because even twin siblings often have different friends (Thorpe & Gardner, 2006), friendship experiences may be an important source of nonshared environmental influence.

In principle, quantitative studies treat all three variance components as unmeasured (latent) variables whose relative influence is statistically estimated without directly assessing study participants' DNA or measuring their environmental experiences. Nevertheless, many quantitative genetic studies do include directly measured features of the environment. This makes it possible to explicitly model rGE or GxE involving specific environmental variables of interest using different statistical procedures such as ACE modeling, Defries-Fulker regression, or genetic risk index modeling (Brendgen et al., 2012; DiLalla & DiLalla, 2018; Pennington et al., 2009). However, a limitation of quantitative genetic studies concerns the interpretation of the estimated effects. For instance, while differences between genetically identical siblings may reflect distinct environmental experiences, they also may be due to measurement error or potential random variations at the molecular level (Neale & Cardon, 1992; Tikhodeyev & Shcherbakova, 2019). Both of these processes can lead to inflated estimates of nonshared environmental effects, which may result in an underestimation of shared environmental and/or genetic influences. Moreover, quantitative genetic designs can only provide a general estimate of heritability, which not only reflects the overall contribution of genes but also of any potential non-measured gene-environment correlations and gene-environment interactions, including heritable epigenetic processes (see chapter "School Social Relations and Child Development: Gene-Environment Interplay", this volume). Hence, quantitative genetic research does not inform us about the specific genes (nor the specific environmental factors, unless directly measured) that contribute to child development. Such information can only be obtained from molecular genetic studies that directly collect and analyze biological material.

Molecular Genetic Studies

The decoding of the human genome and subsequent technological development to easily collect and analyze DNA at falling costs have facilitated the inclusion of molecular information in research. Two methodological streams have dominated the integration of molecular genetics in child development research – candidate gene and genome-wide association studies. **Candidate gene studies** focus on specific sequence variants (polymorphisms) for which there is a plausible biological connection to the phenotype under examination (e.g., aggression) and examine their correlation with individual variance in that phenotype. Selection of candidate variants usually focus on functional sequences that alter the function of a gene (Albert, 2011). These include binary single nucleotide polymorphisms (SNPs), which are single base variation at a specific point in the genome, or variable number tandem repeat polymorphisms (VNTR), which occur in different lengths and typically have more than two variants. Candidate gene studies are based on the assumption that gene function differs for carriers of variants of a SNP or VNTR, resulting in differences in the expression of the trait hypothesized to be related to the respective gene. Next to functioning as main predictors, candidate polymorphisms also have been examined in interplay with environmental risk, both in gene-by-environment interaction studies and in gene-environment correlation models (for detailed overviews, see Halldorsdottir and Binder (2017), Thomas et al. (2015)).

The second stream of molecular genetic research on child development and maladjustment links genetic variants and phenotypes using a data-driven approach in **genome-wide association studies (GWAS)**. Current genotyping chips are able to store information of far over a million different genetic variants, for which statistical associations with the phenotype of interest are calculated. GWAS requires correction for multiple comparisons across a huge number of tests (i.e., one per SNP), necessitating very large samples to achieve gene-wide and genome-wide significance.

Both approaches have initially been met with great excitement, but limitations are recognized now: with respect to candidate studies, it is unlikely that a single polymorphism – of which a human has millions – should have a sizable effect on a complex behavior or disorder. However, because early findings have rarely been replicated, the observed associations between specific polymorphisms and outcomes may have been chance findings. With respect to GWAS, underpowered studies have been named as one of the reasons why initially hardly any genome-wide hits were found for complex traits. Although larger samples are achieved through GWAS consortia, replicated genome-wide SNP-phenotype associations are still rare, especially for complex phenotypes, such as aggressive behavior and emotional symptoms that are of interest to child development researchers. Consequently, in our review of the literature, we emphasize replicated findings and meta-analyses instead of discussing candidate gene studies for which no replications have been published and give preference to well-powered genome-wide association research. We also include recent work using **polygenic scores**, which circumvent many of the limitations of candidate gene research.

For the latter approach, information on single SNPs is combined into polygenic scores (PGS, sometimes polygenic risk scores, PRS). PGS/PRS have initially been created by combining several polymorphisms that were considered to have a meaningful additive function (Belsky & Pluess, 2009; Stavrakakis et al., 2012) or, more recently, by combining coefficients of all SNP-phenotype associations with the phenotype above a certain p-threshold as yielded in a GWAS (e.g., $p < 0.05$, $p < 0.1$, $p < 0.5$). Each individual in a dataset is assigned a PGS which expresses how many alleles related to the trait of interest the individual carries (see details in Belsky and Harden (2019)).

Since PGS are derived from GWAS summary statistics, their predictive power and precision depend on effect sizes of the SNPs that are used, as well as on the sample size of the GWAS. As an example, one of the first GWAS-based PGS was for *years of education* as proxy of educational attainment and cognitive ability, based on a sample of over 125,000 individuals (Rietveld et al., 2013). This PGS has been updated using a GWAS of almost 330,000 individuals (Okbay et al., 2016) and again using a GWAS of approximately 1.1 million individuals (Lee et al., 2018). With these updates, the amount of variance explained in educational attainment has increased from 2% to over 10%. PGS include additive genetic effects only but do not account for gene-gene interaction, and the amount of variance they currently explain is still substantially lower than heritability estimates derived from quantitative genetic studies. As such, PGS constitute a means to control for some genetic influence on a trait but do not contain all genetic influence. PGS are nevertheless highly useful for molecular genetic research into children's psychological and academic adjustment, because a PGS based on a GWAS on, for instance, educational attainment conducted on a specific sample can be used to estimate the amount of variance explained by genetic influence in educational attainment in a different sample. Moreover, PGS allow for smaller sample sizes than GWAS and can be included in prediction models for other phenotypes that are presumed or have been shown to be related to the phenotype on which the GWAS was conducted (Wertz et al., 2018).

GWAS summary statistics are available for various phenotypes with potential relevance to child maladjustment, including internalizing and externalizing disorders (Benke et al., 2014; Howard et al., 2018; Pappa et al., 2015; Tielbeek et al., 2012).

Gene-Environment Interplay: Evidence from Quantitative Genetic Studies

Compared with adoption or twin studies examining gene-environment interplay with respect to family-related risk factors or stressful life events, studies focusing on the school social environment are still few in number. Nevertheless, several quantitative genetic studies have provided increasing evidence that experiences with teachers and peers may be involved in both rGE and GxE. The following sections offer a brief overview of these research findings.

Quantitative Genetic Studies of Externalizing Problems

While externalizing problems encompass a variety of symptoms – including aggressive, oppositional, hyperactive, or delinquent behavior – research on the interplay between genetic factors and the school environment has so far mostly focused on aggression. Studies on the heritability of aggression in humans suggested that around 50% of the variance of aggression is determined by genetic factors (for a review, see Tuvblad and Baker (2011)). Even stronger genetic effects were found in a meta-analysis of 103 twin and adoption studies that only included child and adolescent samples, accounting for 65% of aggressive behavior (Burt, 2009). Nonshared environmental factors explain most of the remaining variance, whereas shared environmental influences only play a small role. What's more, longitudinal research suggests that the high stability of aggressive behavior from childhood to adolescence is mainly due to common underlying genetic influences (Porsch et al., 2016). However, as already mentioned, estimation of genetic (and of shared environmental and nonshared environmental) effects in the standard ACE model used in quantitative genetic studies assumes that there is no gene-environment interplay. The variance attributable to any potential rGE or GxE is typically subsumed in the genetic effect estimate. As such, the high level of variance for aggression explained by genetic effects may, to a considerable part, reflect unmeasured rGE or GxE. Findings from quantitative studies that included direct measures of the environment, notably peer victimization, indeed suggest that such experiences may be correlated – as well as interact – with genetic factors to shape children's and adolescents' developmental adjustment.

Gene-Environment Correlations (rGE)

There is considerable evidence that heritable characteristics of the child can elicit positive or negative reactions from the social environment, reflecting evocative rGE. For instance, a meta-analysis based on 32 children-as-twins studies found that 23% of the variance of parental behavior can be explained by genetically influenced child characteristics (Avinun & Knafo, 2013). Importantly, these genetic effects in regard to parenting most likely indicate evocative rGE, because parent-driven effects on parenting – including the effects of the parents' genes – would be part of the shared environment effect (Klahr & Burt, 2014). Further research shows that child aggression in particular – but not rule-breaking behavior – evokes maternal negativity (Klahr et al., 2014).

Whether similarly pervasive child effects are found for teachers is less clear, as the few existing studies have yielded inconsistent results. For instance, in a longitudinal, nationally representative birth cohort of over 1000 British twins (Environmental Risk (E-Risk) Longitudinal Twin Study), a child's challenging behavior (including irritability, hyperactivity, and inattention) assessed at age 5 predicted how often a teacher had to intervene with that same child in the classroom at age 12 (Houts et al., 2010). This predictive link was entirely accounted for by genetic factors related to the child's problem behavior. In contrast, data from the Quebec Newborn Twin Study (QNTS; a sample of over 600 Canadian twin pairs followed prospectively since birth) (Boivin et al., 2012), revealed no effects of a child's heritable characteristics (including aggression) on teachers' perceived relationship quality with that child in Grade 1 (Brendgen et al., 2011). Although methodological differences in regard to the assessed variables and ages may explain these different findings, more research is needed to examine potential rGE linking children's aggression or other externalizing symptoms with their teachers' behavioral reactions.

Genetically influenced aggressive behavior clearly seems to affect how children are perceived and treated by their classmates, however. Evidence to this effect comes, for instance, from the QNTS where genetic effects explained more than 70% of individual differences in teacher-, peer-, and self-rated rejection and victimization by classmates from kindergarten to Grade 4, especially when these peer difficulties persisted over time (e.g., Boivin et al., 2013). Further analyses suggested that, at the beginning of primary school, all of this genetic influence was explained by genetic factors underlying aggressive behavior (Brendgen et al., 2011). Similar findings were reported for peer victimization at age 10 in the British E-Risk sample (Ball et al., 2008). Because individuals are unlikely to deliberately choose to be rejected and victimized, these findings can be interpreted as indicating evocative rather than active rGE. Perhaps even more compelling evidence comes from an observational study of 5-year-old US twins, each of whom was paired with an unfamiliar same-sex peer in a peer-play interaction, thus ruling out passive (parent-influenced) and active (selective) rGE (DiLalla & John, 2014; DiLalla et al., 2015a; DiLalla & DiLalla, 2018). Results showed that children with a strong genetic disposition for aggressive or rule-breaking behavior were more likely to evoke physical or verbal aggression from the play partner. In contrast, those with a strong genetic disposition for prosocial behavior were more often met with friendly behavior from the play partner.

However, the expression of a genetic disposition for problem behavior may not always evoke negative responses from the environment. For instance, the rGE linking aggressive behavior with negative treatment by peers has been shown to vary depending on the prevailing behavioral norms in the classroom (Brendgen et al.,

2013a). Specifically, children with a strong genetic vulnerability for aggression were found to be at greater risk of being victimized by their classmates only when classroom norms indicated disapproval of aggressive behavior. In contrast, children with a genetic disposition for aggression were *less* likely to be victimized than others in classrooms where such behavior was highly accepted. These results underline the importance of considering multiple levels of the social school context to understand how genetic vulnerabilities may affect children's development – a notion that is also emphasized by the gene-environment interactions discussed next.

Gene-Environment Interactions (GxE)

Several studies based on the QNTS provide evidence that social experiences in school affect children's genetic vulnerabilities to influence externalizing problems, specifically aggression. For instance, children were found to more readily express a genetic disposition for physical aggression when behavioral norms in the classroom favored such behavior (Brendgen et al., 2013a). This "facilitation" effect of favorable peer group norms did not apply to relational aggression, however. Relational aggression often involves covert behaviors such as the spreading of malicious rumors about the victim, which makes it difficult to identify the aggressor. Children with a genetic disposition for relational aggression may therefore use this behavior regardless of whether classroom norms are favorable or not.

Negative treatment by classmates may also promote aggressive behavior especially in genetically vulnerable youth. For example, in line with a diathesis-stress process of GxE, victimized girls show a high level of aggression mainly if they have a high genetic risk for being aggressive (Brendgen et al., 2008). In contrast, a high level of peer victimization seems to predict a high level of aggression regardless of genetic susceptibility among boys. This sex-specific pattern of GxE may in part be due to the fact that girls perceive aggressive behavior as less acceptable than boys (Goldstein et al., 2002). Girls may thus react aggressively to harassment by classmates only if they are highly genetically disposed to such behavior. However, findings of a follow-up study suggest that teachers may help lessen the negative consequences of peer victimization for genetically vulnerable children (Brendgen et al., 2011). Specifically, children with a strong genetic risk for aggression were less likely to express this trait when they had a warm and conflict-free relationship with their teacher. This pattern indicates a compensation process of GxE, with a positive social school environment inhibiting the expression of a child's genetic disposition for aggression. These findings have interesting implications for school-based interventions to curb externalizing problems, but replication studies are needed before finite conclusions can be drawn.

Quantitative Genetic Studies of Internalizing Problems

Internalizing problems during middle childhood typically refer to depression and anxiety but sometimes also encompass related phenotypes such as social withdrawal (Rubin et al., 2009; Zahn-Waxler et al., 2000). Heritable effects have been found for all three components, although estimates vary widely across studies depending on measures and raters. Indeed, some studies found no or only small genetic influences on parent- or self-rated depression among children prior to adolescence (Eley et al., 1998; Eley & Stevenson, 1999; Thapar & McGuffin, 1994). In contrast, using peer-rated depression symptoms, another study found that genetic factors explained around 30% of interindividual differences even when twins were evaluated by different peers (Brendgen et al., 2009). There is some indication, however, that childhood depression is less heritable than adolescent depression (Franić et al., 2010; Rice, 2009). More consistent genetic effects, explaining up to 45% of the variance, have been reported for anxiety behavior prior to adolescence (i.e., among 8- to 12-year-olds) (Eley & Stevenson, 1999; Lau et al., 2007; Legrand et al., 1999). Even stronger heritability was found for socially withdrawn behavior, with genetic influences accounting for up to two thirds of interindividual differences among school-aged children and explaining most of the stability of this behavior over time (Hoekstra et al., 2008). The remaining variance of internalizing symptoms in childhood is accounted for both by shared and nonshared environmental experiences, with the latter explaining at least twice as much variance than the former (Burt, 2009).

Gene-Environment Correlations (rGE)

To our knowledge, quantitative genetic models of gene-and-school environment interplay with respect to internalizing problems in middle childhood have so far only been tested in the QNTS. Overall, the available data suggest that – like children with a genetic disposition for aggression – children with a genetic disposition for internalizing problems may be at risk of eliciting negative reactions from their social environment, especially from their classmates. Indeed, a strong genetic disposition for depressive behavior was related to more peer-nominated rejection in kindergarten and self-reported victimization by classmates in Grade 4 (Brendgen et al., 2009, 2013b). Comparable results were reported in regard to children's (as well as adolescents') genetic disposition for anxiety (Brendgen et al., 2014; Guimond et al., 2015). Observational data seem to corroborate this pattern, as kindergarten peers were found to behave more dominantly toward children with a higher genetic disposition for social withdrawal (Guimond et al., 2014). Despite these circumstances, children's genetic vulnerability for internalizing problems seems to be unrelated to the likelihood of having a reciprocal friend of the number of reciprocal friends in the classroom (Brendgen et al., 2013c). It is still unknown whether children at genetic risk for internalizing problems attract negative reactions from teachers. However, behaviors that are indicative of internalizing problems – such as withdrawal and anxiety – generally do not perturb classroom functioning. Findings from nongenetically informed studies suggest that teachers are not annoyed by children with internalizing problems but instead offer particularly high levels of support to these children (Roorda et al., 2013; Rudasill & Rimm-Kaufman, 2009).

Gene-Environment Interactions (GxE)

Although children with a genetic vulnerability for internalizing problems may be at greater risk of being rejected or even victimized by classmates than others, such negative experiences seem to increase internalizing symptoms in most if not all children, regardless of their genetic vulnerability. Thus, data from the QNTS showed that children who were highly rejected by their peers not only exhibited – on average – more depression symptoms than highly popular children, but the former were also more similar to each other with respect to their depression levels than were the latter. Moreover, genetic factors explained a relatively small portion (about 15%) of interindividual differences in depression symptoms among the highly rejected children. In contrast, genetic influences were considerably more important for the development of depression among highly popular children, explaining about 30% of interindividual differences (Brendgen et al., 2009). It thus seems that – at least at school entry – the contribution of genetic factors to the development of internalizing problems is "suppressed" (i.e., rendered less important) when children are exposed to adverse social experiences in school. While in need of replication, this suppression pattern of GxE may be due to the fact that young children, regardless of their genetic disposition to develop internalizing problems, may be less able to cope effectively with negative peer experiences than older children. Indeed, findings that are more in line with a diathesis-stress process of GxE were reported for anxiety symptoms when the children were in Grade 6 (Guimond et al., 2015). Specifically, at very low levels of peer victimization, even genetically vulnerable children were not more anxious than others. In contrast, among frequently victimized youths, it was the genetically vulnerable individuals who showed the highest levels of anxiety.

Importantly, however, whether or not these processes of GxE occur may itself depends on other features of the children's social environment. For instance, the previously mentioned diathesis-stress process linking peer victimization and anxiety symptoms was only observed when the teacher had low perceived self-efficacy in handling bullying situations or when anti-bullying classroom rules were absent or rarely enforced (Guimond et al., 2015). The crucial role of social support for preventing the development of internalizing problems was demonstrated further when analyzing data collected in Grade 4. The results showed that genetically vulnerable children were less likely to manifest depression symptoms when they had at least one close reciprocal friend in the classroom (Brendgen et al., 2013b). Although replications are needed, these studies offer a first indication of the synergistic interplay between heritable characteristics and multiple aspects of the social school context in influencing children's emotional and behavioral adjustment.

Gene-Environment Interplay: Evidence from Molecular Genetic Studies

Quantitative genetic studies provide important insights about the contributions of genetic and environmental factors to interindividual differences in children's developmental adjustment, but they cannot inform about specific genes at play. Unfortunately, compared to quantitative genetics, there is a dearth of molecular genetic studies that include measures of the school environment as predictors of maladjustment. Indeed, a meta-analysis of gene-environment interaction studies on psychopathology among children showed that most studies examined the environment provided by parents, not the one provided by schools (Pinto et al., 2015). In the following section, we will review the existing studies relevant to internalizing and externalizing problems, paying particular attention to recent work involving innovative molecular genetic methods. We begin with candidate gene, genome-wide association, and polygenic score studies where a direct genetic effect on externalizing and internalizing problems was tested, followed by studies on the interplay between genetic and environmental effects.

Molecular Genetic Studies of Externalizing Problems

Genetic Main Effects

Many candidate gene studies of children's maladjustment examined polymorphisms implicated in brain neurotransmitter systems. Dysregulation of the neurotransmitters dopamine and serotonin is an important factor in several childhood externalizing disorders (including conduct disorder, mood dysregulation, and impulsivity), all of which are components of externalizing behavior. Accordingly, the *5-HTT*LPR gene in the serotonin system and dopamine-related genes *DRD4* and *DAT1* have been studied with respect to externalizing problems (Beitchman et al., 2006; Davidge et al., 2004; Ficks & Waldman, 2014; Lundwall et al., 2017). Similarly, polymorphisms acting on more than one neurotransmitter have been linked to externalizing problems. Examples include *MAOA*, which encodes for the production of an enzyme that plays an important role in neurotransmitter circulation, and the catechol-O-methyltransferase (*COMT*) gene, which codes for the protein that catalyzes dopamine in the brain (Beitchman et al., 2004; Hirata et al., 2013; Sun et al., 2014). However, the variance in externalizing symptoms explained by single genetic variants is small, and reviews and meta-analyses are inconclusive (Weeland et al., 2015). Similarly, GWAS with externalizing phenotypes among children did not support substantial roles of hypothesized genes (Anney et al., 2008; Mick et al., 2011; Pappa et al., 2015).

Importantly, like all complex phenotypes, externalizing problems are highly polygenic, necessitating methods that can account for the accumulation of many tiny effects, such as PGS. For instance, a PGS for ADHD explained irritability in the British ALSPAC sample of over 8000 children and the SAGE sample of children diagnosed with ADHD, but not in the National Child Development Study sample (Riglin et al., 2017). A PGS for ADHD also did not predict externalizing problems in the Generation R study of 2000 children, whereas a PGS for educational attainment predicted variance in aggressive behavior at age 3 and rule-breaking at age 10 (Jansen et al., 2018). A PGS for educational attainment also predicted stable elevated symptoms of conduct problems among 2600 children from the Twins Early Development Study (Hannigan et al., 2018). Finally, a PGS for conduct disorder explained variance in bullying behavior in 6-year-old children (Musci et al., 2018).

Gene-Environment Interplay

Studies examining gene-environment interplay regarding externalizing problems have mostly modeled candidate genes as moderators of risk in the family environment (e.g., Nederhof et al., 2012), whereas research on environmental risk *beyond the family* is rare. The few studies that have been published – on *DRD4* as moderator of links between peer victimization and problem behavior (DiLalla et al., 2015b) and peer social standing and conduct problems (Buil et al., 2015) – warrant replication. Notably, *5-HTT*LPR interacted with teacher-reported peer problems in the prediction of hyperactivity-impulsivity in two independent samples from Norway and the USA (Stenseng et al., 2018). Children who carried the short allele of the *5-HTT*LPR polymorphism and who were not liked, did not get along with, and were teased by their peers were at greater risk for hyperactivity-impulsivity than children who were also exposed to peer problems but did not carry the short allele. Note that this study did not test rGE. It will be a task of future research to explore the interplay between PGS and the social school environment in predicting externalizing problems, as such studies do not exist yet. Generally, studies are needed that elucidate whether and how environmental conditions mediate genetic risk for externalizing maladjustment.

Molecular Genetic Studies of Internalizing Problems

Genetic Main Effects

Just as with studies on externalizing maladjustment, variants related to neurotransmitter systems have also received attention in research on internalizing problems, given their role in emotion regulation (Barzman et al., 2015). However, meta-analyses on direct associations between candidate polymorphisms and internalizing problems failed to find a direct effect on emotional problems in children (Benke et al., 2014; Sallis et al., 2017). As a consequence, acknowledging the polygenic nature of internalizing problems, recent analyses have gone beyond single candidate gene models in research on internalizing maladjustment: for instance, among approximately 2000 children in *Generation R* – a longitudinal study of children and their parents in the Dutch city of Rotterdam – a PGS for educational attainment predicted lower risk of internalizing problems in early childhood only, whereas a PGS for schizophrenia predicted higher risk of internalizing problems across childhood (Jansen et al. 2018). Analyses of over 8600 children from the ALSPAC study and the Dutch Twin Register also yielded an association between a PGS for schizophrenia and internalizing problems in childhood (Nivard et al., 2017). Carriers of fewer education-related alleles were at greater risk of following a trajectory of elevated internalizing and externalizing maladjustment symptoms (Hannigan et al., 2018). In contrast, emotion recognition – a phenotype associated with internalizing problems – was not predicted by PGSs for schizophrenia, bipolar disorder, depression, autism spectrum disorders, or anxiety in the Twins Early Development Study sample of 4000 8-year-old children (Coleman et al., 2017).

Gene-Environment Interplay

Rather than examining *5-HTT*LPR as direct predictor of internalizing maladjustment, several studies conceptualized the polymorphism as moderator of environmental risk, for instance, with respect to prenatal maternal health (Geels et al., 2012), birth weight (Broekman et al., 2011), and parenting (Pauli-Pott et al., 2009). Research beyond the family environment is again scarce, with the exception of a gene-by-environment study on *5-HTT*LPR and bullying victimization conducted in the British E-Risk sample. In that study, frequently victimized children who carried at least one short allele of *5-HTT*LPR were at greater risk for parent-reported concurrent emotional problems compared to carriers of two long alleles (Sugden et al., 2010).

As with externalizing problems, replications of interactions between candidate genes and environmental measures in the prediction of internalizing problems are rare, which means that results may be due to chance. Gene-environment interplay studies involving PGS and social contexts beyond the family have not been published yet, which is unfortunate given recent insights into genetic vulnerability to bullying victimization using PGS for a variety of mental health problems (Schoeler et al., 2019). In that study, the genetic vulnerability for depression was also predictive of bullying victimization – though modestly – suggesting evidence of an evocative rGE; bullying victimization might thus act as environmental mediator of the expression of genetic risk for internalizing problems.

Conclusion and Future Directions

Genetically informed studies offer important insights into the interplay between environmental experiences and genetic dispositions that shape children's developmental adjustment. Despite this advantage, there is still very little genetically informed research that focuses specifically on children's social environment in school. Nevertheless, the few existing findings – particularly those emerging from quantitative genetic research – are quite convergent: genetic liabilities related to internalizing and externalizing symptoms not only play a role in how children and adolescents are perceived and treated by their classmates and teachers, but the social school environment also often interacts with genetic risks to foster mental health problems during childhood. The latter findings have important implications for school-based prevention programs, especially those targeting externalizing problems, as they suggest that genetically vulnerable children might indeed benefit most from modifications of social environmental features of the classroom (e.g., classroom behavior norms, peer interactions, or teacher-child interactions). In other words, while universal anti-bullying programs aimed at improving such key features of the classroom or school environment may yield some positive effects for all children, they may translate into the greatest improvements for those who are most at risk for developing problems. Prevention studies combining strong experimental designs with genetic information are needed to test this hypothesis. If convergent results are found, it would suggest that – depending on the targeted phenotype and environmental features – genetically vulnerable children may not necessarily need specifically tailored prevention approaches, resulting in reduced cost for schools.

Many other questions also remain. One concerns the need for replication of existing findings in different samples – which is essential to increase confidence in study results (Schmidt, 2009). Future genetically informed research should also include other measures of the social school environment that have been linked with children's developmental adjustment (e.g., classroom climate, ethnic composition of the classroom, Fraser, 2012; Graham et al., 2009) and other outcomes, notably children's academic motivation and performance. In addition, genetically informed studies should employ longitudinal designs to explore developmental changes in the interplay between children's genetic dispositions and their social experiences in school. Future research should also aim to integrate measures of the social school environment with epigenetic measures (i.e., individual differences in gene activation due to submolecular modifications to the gene). Indeed, even individuals with the same genotype may show considerable phenotypic differences, depending on whether certain genes are "switched" on or off and which other genes are activated. Although the epigenome is laid down during gestation, it can be modified by later postnatal environmental (including social) influences (Dunn et al., 2019; Marini et al., 2020). Unfortunately, the few published studies with human participants – notably those focusing on victimization in middle childhood and adolescence as the environmental measure – have not always produced consistent results. Specifically, whereas exposure to peer victimization was linked to higher levels of DNA methylation in some studies (Efstathopoulos et al., 2018; Ouellet-Morin et al., 2013), other research found no association (Marzi et al., 2018). Moreover, recent longitudinal data have called into question whether the epigenome is malleable to environmental experiences after infancy or whether sensitive periods are restricted to the early years (Dunn et al., 2019). Future methodological and technical advances in data assaying may eventually offer a more nuanced picture of how environmental experiences – including those in school – may affect gene expression and, thus, children's development.

To obtain robust results, future genetically informed studies should ideally be based on much larger samples than usually employed in child development research. While this also applies to quantitative genetic designs, it is particularly true for molecular genetic research that aims to detect small effects in genome-wide association studies and replicate candidate gene associations in independent samples. Sufficiently powered quantitative and molecular genetic studies are also required to examine potential sex differences. Consortia have been formed to combine samples to increase power in both quantitative and molecular genetic studies, but only few focus on children, such as the previously mentioned EAGLE consortium for GWAS studies or the ACTION project for twin studies (EU-Aggression in Children: Unravelling Gene-Environment Interplay to Inform Treatment and InterventiON strategies; http://www. action-euproject.eu/). Findings from quantitative genetic studies may also help inform the selection of environmental variables for testing rGE and GxE in molecular genetic research. As summary statistics from increasingly well-powered GWAS become available, it will be fascinating to follow how GWAS-based polygenic scores for closely related traits such as aggression, depression, and educational attainment will perform in explaining variance in children's psychological and academic maladjustment, alone and in interplay with environmental factors, including the social school environment. As noted by Hatemi and colleagues (Hatemi et al., 2011, p. 81): "Only by considering both the environmental and genetic sources of individual differences can we gain a deep understanding of behavior. The more we learn about how genes lead us into environments, affect our interpretations of the exogenous environments we encounter, and how our social environments may change our genetic expression, the more we can contribute to the discipline at large about which environments matter and why."

References

Albert, P. R. (2011). What is a functional genetic polymorphism? Defining classes of functionality. *Journal of Psychiatry & Neuroscience: JPN, 36*(6), 363–365. https://doi.org/10.1503/jpn.110137

Anney, R. J., Lasky-Su, J., Ó'Dúshláine, C., Kenny, E., Neale, B. M., Mulligan, A., et al. (2008). Conduct disorder and ADHD: Evaluation of conduct problems as a categorical and quantitative trait in the international multicentre ADHD genetics study. *American Journal of Medical Genetics Part B: Neuropsychiatric Genetics, 147*(8), 1369–1378.

Avinun, R., & Knafo, A. (2013). Parenting as a reaction evoked by children's genotype: A meta-analysis of children-as-twins studies. *Personality and Social Psychology Review, 18*(1), 87–102. https://doi.org/10.1177/1088868313498308

Ball, H. A., Arseneault, L., Taylor, A., Maughan, B., Caspi, A., & Moffitt, T. E. (2008). Genetic and environmental influences on victims, bullies and bully-victims in childhood. *Journal of Child Psychology and Psychiatry, 49*(1), 104–112. https://doi.org/10.1111/j.1469-7610.2007.01821.x

Barzman, D., Geise, C., & Lin, P.-I. (2015). Review of the genetic basis of emotion dysregulation in children and adolescents. *World Journal of Psychiatry, 5*(1), 112–117. https://doi.org/10.5498/wjp.v5.i1.112

Beitchman, J. H., Mik, H. M., Ehtesham, S., Douglas, L., & Kennedy, J. L. (2004). MAOA and persistent, pervasive childhood aggression. *Molecular Psychiatry, 9*, 546–547. https://doi.org/10.1038/sj.mp.4001492

Beitchman, J., Baldassarra, L., Mik, H., Luca, V. D., King, N., Bender, D., … Kennedy, J. (2006). Serotonin transporter polymorphisms and persistent, pervasive childhood aggression. *American Journal of Psychiatry, 163*(6), 1103–1105.

Belsky, D. W., & Harden, K. P. (2019). Phenotypic annotation: Using polygenic scores to translate discoveries from genome-wide association studies from the top down. *Current Directions in Psychological Science, 28*(1), 82–90. https://doi.org/10.1177/0963721418807729

Belsky, J., & Pluess, M. (2009). Beyond diathesis stress: Differential susceptibility to environmental influences. *Psychological Bulletin, 135*(6), 885–908.

Benke, K. S., Nivard, M. G., Velders, F. P., Walters, R. K., Pappa, I., Scheet, P. A., … Middeldorp, C. M. (2014). A genome-wide association meta-analysis of preschool internalizing problems. *Journal of the American Academy of Child & Adolescent Psychiatry, 53*(6), 667–676. e667. https://doi.org/10.1016/j.jaac.2013.12.028

Boivin, M., Vitaro, F., & Poulin, F. (2005). Peer relationships and the development of aggressive behavior in early childhood. In R. E. Tremblay, W. W. Hartup, & J. Archer (Eds.), *Developmental origins of aggression* (pp. 376–397). Guilford Press.

Boivin, M., Brendgen, M., Dionne, G., Dubois, L., Pérusse, D., Robaey, P., . . . Vitaro, F. (2012). The Quebec newborn twin study into adolescence: 15 Years later. *Twin research and human genetics, FirstView*, pp. 1–6. https://doi.org/10.1017/thg.2012.129

Boivin, M., Brendgen, M., Vitaro, F., Dionne, G., Girard, A., Pérusse, D., & Tremblay, R. E. (2013). Strong genetic contribution to peer relationship difficulties at school entry: Findings from a longitudinal twin study. *Child Development, 84*(3), 1098–1114. https://doi.org/10.1111/cdev.12019

Brendgen, M., Boivin, M., Vitaro, F., Dionne, G., Girard, A., & Pérusse, D. (2008). Gene-environment interactions between peer victimization and child aggression. *Development and Psychopathology, 20*, 455–471.

Brendgen, M., Vitaro, F., Boivin, M., Girard, A., Bukowski, W. M., Dionne, G., … Pérusse, D. (2009). Gene-environment linkages between peer rejection and depressive symptoms in children. *Journal of Child Psychology and Psychiatry, 50*(8), 1009–1017.

Brendgen, M., Boivin, M., Dionne, G., Barker, E. D., Vitaro, F., Girard, A., … Pérusse, D. (2011). Gene–environment processes linking aggression, peer victimization, and the teacher–child relationship. *Child Development, 82*(6), 2021–2036. https://doi.org/10.1111/j.1467-8624.2011.01644.x

Brendgen, M., Vitaro, F., & Girard, A. (2012). Evaluating gene-environment interplay. In *Handbook of development research methods* (pp. 687–705).

Brendgen, M., Girard, A., Vitaro, F., Dionne, G., & Boivin, M. (2013a). Gene-environment correlation linking aggression and peer victimization: Do classroom behavioral norms matter? *Journal of Abnormal Child Psychology*, 1–13. https://doi.org/10.1007/s10802-013-9807-z

Brendgen, M., Vitaro, F., Barker, E. D., Girard, A., Dionne, G., Tremblay, R. E., & Boivin, M. (2013b). Do other people's plights matter? A genetically informed twin study of the role of social context in the link between peer victimization and children's aggression and depression symptoms. *Developmental Psychology, 49*(2), 327–340. https://doi.org/10.1037/a0025665

Brendgen, M., Vitaro, F., Bukowski, W. M., Dionne, G., Tremblay, R. E., & Boivin, M. (2013c). Can friends protect genetically vulnerable children from depression? *Development and Psychopathology, 25*(2), 277–289. https://doi.org/10.1017/S0954579412001058

Brendgen, M., Girard, A., Vitaro, F., Dionne, G., & Boivin, M. (2014). The dark side of friends: A genetically informed study of victimization within early adolescents' friendships. *Journal of Clinical Child & Adolescent Psychology*, 1–15. https://doi.org/10.1080/15374416.2013.873984

Broekman, B. F. P., Chan, Y. H., Goh, L., Fung, D., Gluckman, P. D., Saw, S. M., & Meaney, M. J. (2011). Influence of birth weight on internalizing traits modulated by serotonergic genes. *Pediatrics, 128*(5), e1250–e1258. https://doi.org/10.1542/peds.2010-3714

Buil, J. M., Koot, H., Olthof, T., Nelson, K., & van Lier, P. C. (2015). DRD4 genotype and the developmental link of peer social preference with conduct problems and prosocial behavior across ages 9–12 years. *Journal of youth and adolescence, 44*(7), 1360–1378. https://doi.org/10.1007/s10964-015-0289-x

Burt, S. A. (2009). Rethinking environmental contributions to child and adolescent psychopathology: A meta-analysis of shared environmental influences. *Psychological Bulletin, 135*(4), 608–637. https://doi.org/10.1037/a0015702

Coleman, J. R. I., Lester, K. J., Keers, R., Munafò, M. R., Breen, G., & Eley, T. C. (2017). Genome-wide association study of facial emotion recognition in children and association with polygenic risk for mental health disorders. *American Journal of Medical Genetics Part B: Neuropsychiatric Genetics, 174*(7), 701–711. https://doi.org/10.1002/ajmg.b.32558

Conger, K. J., & Conger, R. D. (1994). Differential parenting and change in sibling differences in delinquency. *Journal of Family Psychology, 8*, 287–302.

Costa, A. M., Breitenfeld, L., Silva, A. J., Pereira, A., Izquierdo, M., & Marques, M. C. (2012). Genetic inheritance effects on endurance and muscle strength. *Sports Medicine, 42*(6), 449–458. https://doi.org/10.2165/11650560-000000000-00000

Davidge, K. M., Atkinson, L., Douglas, L., Lee, V., Shapiro, S., Kennedy, J. L., & Beitchman, J. H. (2004). Association of the serotonin transporter and 5ht1dβ receptor genes with extreme, persistent and pervasive aggressive behaviour in children. *Psychiatric Genetics, 14*(3), 143–146.

DiLalla, L. F., & DiLalla, D. L. (2018). Gene–environment correlations affecting children's early rule-breaking and aggressive play behaviors. *Twin Research and Human Genetics, 21*(4), 285–288. https://doi.org/10.1017/thg.2018.30

DiLalla, L., & John, S. G. (2014). Genetic and behavioral influences on received aggression during observed play among unfamiliar preschool-aged peers. *Merrill-Palmer Quarterly, 60*(2), 168–192. https://doi.org/10.13110/merrpalmquar1982.60.2.0168

DiLalla, L. F., Bersted, K., & John, S. (2015a). Peer victimization and DRD4 genotype influence problem behaviors in young children. *Journal of Youth and Adolescence, 44*(8), 1478–1493. https://doi.org/10.1007/s10964-015-0282-4

DiLalla, L. F., Bersted, K., & John, S. G. (2015b). Evidence of reactive gene-environment correlation in preschoolers' prosocial play with unfamiliar peers. *Developmental Psychology, 51*(10), 1464–1475. https://doi.org/10.1037/dev0000047

Dunn, J., & Plomin, R. (1990). *Separate lives: Why siblings are so different*. Basic Books.

Dunn, E. C., Soare, T. W., Zhu, Y., Simpkin, A. J., Suderman, M. J., Klengel, T., … Relton, C. L. (2019). Sensitive periods for the effect of childhood adversity on DNA methylation: Results from a prospective, longitudinal study. *Biological psychiatry, 85*(10), 838–849. https://doi.org/10.1016/j.biopsych.2018.12.023

Efstathopoulos, P., Andersson, F., Melas, P. A., Yang, L. L., Villaescusa, J. C., Rüegg, J., … Lavebratt, C. (2018). NR3C1 hypermethylation in depressed and bullied adolescents. *Translational Psychiatry, 8*(1), 121. https://doi.org/10.1038/s41398-018-0169-8

Eley, T. C., & Stevenson, J. (1999). Exploring the covariation between anxiety and depression symptoms: a genetic analysis of the effects of age and sex. *Journal of Child Psychology and Psychiatry, 40*(8), 1273–1282. https://doi.org/10.1111/1469-7610.00543

Eley, T. C., Deater-Deckard, K., Fombonne, E., Fulker, D. W., & Plomin, R. (1998). An adoption study of depressive symptoms in middle childhood. *Journal of Child Psychology and Psychiatry, 39*(3), 337–345. https://doi.org/10.1111/1469-7610.00329

Ficks, C. A., & Waldman, I. D. (2014). Candidate genes for aggression and antisocial behavior: a meta-analysis of association studies of the 5HTTLPR and MAOA-uVNTR. *Behavior Genetics, 44*(5), 427–444.

Franić, S., Middeldorp, C. M., Dolan, C. V., Ligthart, L., & Boomsma, D. I. (2010). Childhood and adolescent anxiety and depression: Beyond heritability. *Journal of the American Academy of Child & Adolescent Psychiatry, 49*(8), 820–829.

Fraser, B. J. (2012). *Classroom environment* (Vol. 234). Routledge.

Geels, L. M., Groen-Blokhuis, M. M., van Beijsterveldt, C. E. M., Vink, J. M., Middeldorp, C. M., Bartels, M., … Boomsma, D. I. (2012). Maternal prenatal smoking and offspring emotional problems: No moderating effect of maternal or child 5-HTTLPR genotype. *American Journal of Medical Genetics Part B: Neuropsychiatric Genetics, 159B*(8), 1009–1012. https://doi.org/10.1002/ajmg.b.32110

Goldstein, S. E., Tisak, M. S., & Boxer, P. (2002). Preschoolers' normative and prescriptive judgments about relational and overt aggression. *Early Education and Development, 13*(1), 23–39.

Graham, S., Bellmore, A., Nishina, A., & Juvonen, J. (2009). "it must be me": Ethnic diversity and attributions for peer victimization in middle school. *Journal of Youth and Adolescence, 38*(4), 487–499.

Guimond, F.-A., Brendgen, M., Vitaro, F., Forget-Dubois, N., Dionne, G., Tremblay, R. E., & Boivin, M. (2014). Gene–environment interplay in the link of friends' and nonfriends' behaviors with children's social reticence in a competitive situation. *Developmental Psychology, 50*(3), 956.

Guimond, F.-A., Brendgen, M., Vitaro, F., Dionne, G., & Boivin, M. (2015). Peer victimization and anxiety in genetically vulnerable youth: The protective roles of teachers' self-efficacy and anti-bullying classroom rules. *Journal of Abnormal Child Psychology, 43*(6), 1095–1106. https://doi.org/10.1007/s10802-015-0001-3

Halldorsdottir, T., & Binder, E. B. (2017). Gene environment interactions: from molecular mechanisms to behavior. *Annual Review of Psychology, 68*, 215–241.

Hannigan, L. J., Pingault, J.-B., Krapohl, E., McAdams, T. A., Rijsdijk, F. V., & Eley, T. C. (2018). Genetics of co-developing conduct and emotional problems during childhood and adolescence. *Nature Human Behaviour, 2*(7), 514. https://doi.org/10.1038/s41562-018-0373-9

Hatemi, P. K., Dawes, C. T., Frost-Keller, A., Settle, J. E., & Verhulst, B. (2011). Integrating social science and genetics: News from the political front. *Biodemography and Social Biology, 57*(1), 67–87. https://doi.org/10.1080/19485565.2011.568276

Hirata, Y., Zai, C. C., Nowrouzi, B., Beitchman, J. H., & Kennedy, J. L. (2013). Study of the Catechol-O-Methyltransferase (COMT) gene with high aggression in children. *Aggressive Behavior, 39*(1), 45–51.

Hoekstra, R. A., Bartels, M., Hudziak, J. J., Van Beijsterveldt, T. C. E. M., & Boomsma, D. I. (2008). Genetic and environmental influences on the stability of withdrawn behavior in children: A longitudinal, multi-informant twin study. *Behavior Genetics, 38*(5), 447–461. https://doi.org/10.1007/s10519-008-9213-4

Houts, R. M., Caspi, A., Pianta, R. C., Arseneault, L., & Moffitt, T. E. (2010). The challenging pupil in the classroom: The effect of the child on the teacher. *Psychological Science*. https://doi.org/10.1177/0956797610388047

Howard, D. M., Adams, M. J., Shirali, M., Clarke, T.-K., Marioni, R. E., Davies, G., … McIntosh, A. M. (2018). Genome-wide association study of depression phenotypes in UK Biobank identifies variants in excitatory synaptic pathways. *Nature Communications, 9*(1), 1470. https://doi.org/10.1038/s41467-018-03819-3

Jansen, P. R., Polderman, T. J. C., Bolhuis, K., van der Ende, J., Jaddoe, V. W. V., Verhulst, F. C., … Tiemeier, H. (2018). Polygenic scores for schizophrenia and educational attainment

are associated with behavioural problems in early childhood in the general population. *Journal of Child Psychology and Psychiatry, 59*(1), 39–47. https://doi.org/10.1111/jcpp.12759
Kindermann, T. A. (2011). Commentary: The invisible hand of the teacher. *Journal of Applied Developmental Psychology, 32*(5), 304–308. https://doi.org/10.1016/j.appdev.2011.04.005
Klahr, A. M., & Burt, S. A. (2014). Elucidating the etiology of individual differences in parenting: A meta-analysis of behavioral genetic research. *Psychological Bulletin, 140*(2), 544–586. https://doi.org/10.1037/a0034205
Klahr, A. M., Klump, K. L., & Burt, S. A. (2014). The etiology of the association between child antisocial behavior and maternal negativity varies across aggressive and non-aggressive rule-breaking forms of antisocial behavior. *Journal of Abnormal Child Psychology, 42*(8), 1299–1311. https://doi.org/10.1007/s10802-014-9886-5
Kovas, Y., Haworth, C. M. A., Dale, P. S., Plomin, R., Weinberg, R. A., Thomson, J. M., & Fischer, K. W. (2007). The genetic and environmental origins of learning abilities and disabilities in the early school years. *Monographs of the Society for Research in Child Development, 72*(3), i–156.
Lau, J. Y. F., Gregory, A. M., Goldwin, M. A., Pine, D. S., & Eley, T. C. (2007). Assessing gene-environment interactions on anxiety symptom subtypes across childhood and adolescence. *Development and Psychopathology, 19*(04), 1129–1146. https://doi.org/10.1017/S0954579407000582
Lee, J. J., Wedow, R., Okbay, A., Kong, E., Maghzian, O., Zacher, M., … Cesarini, D. (2018). Gene discovery and polygenic prediction from a genome-wide association study of educational attainment in 1.1 million individuals. *Nature Genetics, 50*(8), 1112–1121. https://doi.org/10.1038/s41588-018-0147-3
Legrand, L. N., McGue, M., & Iacono, W. G. (1999). A twin study of state and trait anxiety in childhood and adolescence. *Journal of Child Psychology and Psychiatry, 40*(6), 953–958. https://doi.org/10.1111/1469-7610.00512
Lundwall, R. A., Sgro, J., & Wade, T. (2017). SLC6A3 is associated with relational aggression in children. *Journal of Individual Differences, 38*(4), 220.
Marini, S., Davis, K. A., Soare, T. W., Zhu, Y., Suderman, M. J., Simpkin, A. J., ... & Dunn, E. C. (2020). Adversity exposure during sensitive periods predicts accelerated epigenetic aging in children. *Psychoneuroendocrinology, 113*, 104484.
Marzi, S. J., Sugden, K., Arseneault, L., Belsky, D. W., Burrage, J., Corcoran, D. L., … Moffitt, T. E. (2018). Analysis of DNA methylation in young people: limited evidence for an association between victimization stress and epigenetic variation in blood. *American Journal of Psychiatry, 175*(6), 517–529.
Mick, E., McGough, J., Loo, S., Doyle, A. E., Wozniak, J., Wilens, T. E., … Faraone, S. V. (2011). Genome-wide association study of the child behavior checklist dysregulation profile. *Journal of the American Academy of Child & Adolescent Psychiatry, 50*(8), 807–817. e808. https://doi.org/10.1016/j.jaac.2011.05.001
Musci, R. J., Bettencourt, A. F., Sisto, D., Maher, B., Uhl, G., Ialongo, N., & Bradshaw, C. P. (2018). Evaluating the genetic susceptibility to peer reported bullying behaviors. *Psychiatry Research, 263*, 193–198. https://doi.org/10.1016/j.psychres.2018.03.016
Neale, M. C., & Cardon, L. R. (1992). *Methodology for genetic studies of twins and families*. Kluwer Academic.
Nederhof, E., Belsky, J., Ormel, J., & Oldehinkel, A. J. (2012). Effects of divorce on Dutch boys' and girls' externalizing behavior in Gene × Environment perspective: Diathesis stress or differential susceptibility in the Dutch Tracking Adolescents' Individual Lives Survey study? *Development and Psychopathology, 24*(3), 929–939. https://doi.org/10.1017/S0954579412000454
Nivard, M. G., Gage, S. H., Hottenga, J. J., van Beijsterveldt, C. E. M., Abdellaoui, A., Bartels, M., … Middeldorp, C. M. (2017). Genetic overlap between schizophrenia and developmental psychopathology: Longitudinal and multivariate polygenic risk prediction of common psychiatric traits during development. *Schizophrenia Bulletin, 43*(6), 1197–1207. https://doi.org/10.1093/schbul/sbx031

Okbay, A., Beauchamp, J. P., Fontana, M. A., Lee, J. J., Pers, T. H., Rietveld, C. A., … Benjamin, D. J. (2016). Genome-wide association study identifies 74 loci associated with educational attainment. *Nature, 533*(7604), 539–542. https://doi.org/10.1038/nature17671

Ouellet-Morin, I., Wong, C., Danese, A., Pariante, C., Papadopoulos, A., Mill, J., & Arseneault, L. (2013). Increased serotonin transporter gene (SERT) DNA methylation is associated with bullying victimization and blunted cortisol response to stress in childhood: A longitudinal study of discordant monozygotic twins. *Psychological Medicine, 43*(09), 1813–1823. https://doi.org/10.1017/S0033291712002784

Pappa, I., St Pourcain, B., Benke, K., Cavadino, A., Hakulinen, C., Nivard, M. G., et al. (2015). A genome-wide approach to children's aggressive behavior: The EAGLE consortium. *American Journal of Medical Genetics Part B: Neuropsychiatric Genetics*, 562–572. https://doi.org/10.1002/ajmg.b.32333

Pauli-Pott, U., Friedl, S., Hinney, A., & Hebebrand, J. (2009). Serotonin transporter gene polymorphism (5-HTTLPR), environmental conditions, and developing negative emotionality and fear in early childhood. *Journal of Neural Transmission, 116*(4), 503–512.

Pennington, B. F., McGrath, L. M., Rosenberg, J., Barnard, H., Smith, S. D., Willcutt, E. G., … Olson, R. K. (2009). Gene × environment interactions in reading disability and attention-deficit/hyperactivity disorder. *Developmental Psychology, 45*(1), 77–89. https://doi.org/10.1037/a0014549

Pinto, R. Q., Soares, I., Carvalho-Correia, E., & Mesquita, A. R. (2015). Gene–environment interactions in psychopathology throughout early childhood: A systematic review. *Psychiatric Genetics, 25*(6), 223–233. https://doi.org/10.1097/YPG.0000000000000106

Plomin, R. (2011). Commentary: Why are children in the same family so different? Non-shared environment three decades later. *International Journal of Epidemiology, 40*(3), 582–592. https://doi.org/10.1093/ije/dyq144

Plomin, R., DeFries, J. C., Knopik, V. S., & Neiderhiser, J. M. (2016). Top 10 replicated findings from behavioral genetics. *Perspectives on Psychological Science, 11*(1), 3–23. https://doi.org/10.1177/1745691615617439

Pluess, M., & Belsky, J. (2013). Vantage sensitivity: Individual differences in response to positive experiences. *Psychological Bulletin, 139*(4), 901–916. https://doi.org/10.1037/a0030196

Polderman, T. J. C., Bartels, M., Verhulst, F. C., Huizink, A. C., van Beijsterveldt, C. E. M., & Boomsma, D. I. (2010). No effect of classroom sharing on educational achievement in twins: A prospective, longitudinal cohort study. *Journal of Epidemiology and Community Health, 64*(01), 36. https://doi.org/10.1136/jech.2009.091629

Porsch, R. M., Middeldorp, C. M., Cherny, S. S., Krapohl, E., van Beijsterveldt, C. E. M., Loukola, A., … Bartels, M. (2016). Longitudinal heritability of childhood aggression. *American Journal of Medical Genetics Part B: Neuropsychiatric Genetics, 171*(5), 697–707. https://doi.org/10.1002/ajmg.b.32420

Rende, R., & Plomin, R. (1992). Diathesis-stress models of psychopathology: A quantitative genetic perspective. *Applied and Preventive Psychology, 1*(4), 177–182. https://doi.org/10.1016/S0962-1849(05)80123-4

Rice, F. (2009). The genetics of depression in childhood and adolescence. *Current Psychiatry Reports, 11*(2), 167–173. https://doi.org/10.1007/s11920-009-0026-9

Rietveld, C. A., Medland, S. E., Derringer, J., Yang, J., Esko, T., Martin, N. W., … Koellinger, P. D. (2013). GWAS of 126,559 individuals identifies genetic variants associated with educational attainment. *Science, 340*(6139), 1467–1471. https://doi.org/10.1126/science.1235488

Riglin, L., Eyre, O., Cooper, M., Collishaw, S., Martin, J., Langley, K., … Thapar, A. (2017). Investigating the genetic underpinnings of early-life irritability. *Translational Psychiatry, 7*(9), e1241. https://doi.org/10.1038/tp.2017.212

Roorda, D. L., Koomen, H. M. Y., Spilt, J. L., Thijs, J. T., & Oort, F. J. (2013). Interpersonal behaviors and complementarity in interactions between teachers and kindergartners with a variety of externalizing and internalizing behaviors. *Journal of School Psychology, 51*(1), 143–158. https://doi.org/10.1016/j.jsp.2012.12.001

Røysamb, E., & Tambs, K. (2016). The beauty, logic and limitations of twin studies. *Norsk Epidemiologi, 26*(1/2), 35–46.

Rubin, K. H., Coplan, R. J., & Bowker, J. C. (2009). *Social withdrawal in childhood* (Vol. 60, pp. 141–171).

Rudasill, K. M., & Rimm-Kaufman, S. E. (2009). Teacher–child relationship quality: The roles of child temperament and teacher–child interactions. *Early Childhood Research Quarterly, 24*(2), 107–120. https://doi.org/10.1016/j.ecresq.2008.12.003

Rutter, M., Moffitt, T. E., & Caspi, A. (2006). Gene–environment interplay and psychopathology: Multiple varieties but real effects. *Journal of Child Psychology and Psychiatry, 47*(3/4), 226–261.

Sallis, H., Evans, J., Wootton, R., Krapohl, E., Oldehinkel, A. J., Davey Smith, G., & Paternoster, L. (2017). Genetics of depressive symptoms in adolescence. *BMC Psychiatry, 17*(321). https://doi.org/10.1186/s12888-017-1484-y

Scarr, S., & McCartney, K. (1983). How people make their own environments: A theory of genotype – Environment effects. *Child Development, 54*(2), 424–435.

Schmidt, S. (2009). Shall we really do it again? The powerful concept of replication is neglected in the social sciences. *Review of General Psychology, 13*(2), 90–100. https://doi.org/10.1037/a0015108

Schoeler, T., Choi, S. W., Dudbridge, F., Baldwin, J., Duncan, L., Cecil, C. M., ... & Pingault, J. B. (2019). Multi–polygenic score approach to identifying individual vulnerabilities associated with the risk of exposure to bullying. *JAMA psychiatry, 76*(7), 730–738.

Shanahan, M., & Hofer, S. (2005). Social context in gene–environment interactions: Retrospect and prospect. *Journal of Gerontology: Series B, 60B*, 65–76.

Stavrakakis, N., Oldehinkel, A. J., Nederhof, E., OudeVoshaar, R. C., Verhulst, F. C., Ormel, J., & Jonge, P. D. (2012). Plasticity genes do not modify associations between physical activity and depressive symptoms. *Health Psychology, 32*(7), 785–792.

Stenseng, F., Li, Z., Belsky, J., Hygen, B. W., Skalicka, V., Guzey, I. C., & Wichstrøm, L. (2018). Peer problems and hyperactivity-impulsivity among Norwegian and American children: The role of *5-HTTLPR*. *Child Development, 89*(2), 509–524. https://doi.org/10.1111/cdev.12766

Sugden, K., Arseneault, L., Harrington, H., Moffitt, T. E., Williams, B., & Caspi, A. (2010). Serotonin transporter gene moderates the development of emotional problems among children following bullying victimization. *Journal of the American Academy of Child and Adolescent Psychiatry, 49*(8), 830–840. https://doi.org/10.1016/j.jaac.2010.01.024

Sun, H., Yuan, F., Shen, X., Xiong, G., & Wu, J. (2014). Role of COMT in ADHD: A systematic meta-analysis. *Molecular Neurobiology, 49*(1), 251–261. https://doi.org/10.1007/s12035-013-8516-5

Thapar, A., & McGuffin, P. (1994). A twin study of depressive symptoms in childhood. *British Journal of Psychiatry, 165*(2), 259–265. https://doi.org/10.1192/bjp.165.2.259

Thomas, M. S. C., Kovas, Y., Meaburn, E. L., & Tolmie, A. (2015). What can the study of genetics offer to educators? *Mind Brain and Education, 9*(2), 72–80. https://doi.org/10.1111/mbe.12077

Thorpe, K., & Gardner, K. (2006). Twins and their friendships: Differences between monozygotic, dizygotic same-sex and dizygotic mixed-sex pairs. *Twin research and human genetics, 9*(1), 155–164.

Tielbeek, J. J., Medland, S. E., Benyamin, B., Byrne, E. M., Heath, A. C., Madden, P. A., … Verweij, K. J. (2012). Unraveling the genetic etiology of adult antisocial behavior: A genome-wide association study. *PLoS ONE, 7*(10), e45086.

Tikhodeyev, O. N., & Shcherbakova, O. V. (2019). The problem of non-shared environment in behavioral genetics. *Behavior Genetics, 49*(3), 259–269. https://doi.org/10.1007/s10519-019-09950-1

Tuvblad, C., & Baker, L. A. (2011). Human aggression across the lifespan. In *Genetic propensities and environmental moderators* (Vol. 75, pp. 171–214).

Weeland, J., Overbeek, G., Castro, B. O., & Matthys, W. (2015). Underlying mechanisms of gene–environment interactions in externalizing behavior: A systematic review and search for theo-

retical mechanisms. *Clinical Child and Family Psychology Review, 18*(4), 413–442. https://doi.org/10.1007/s10567-015-0196-4
Wertz, J., Caspi, A., Belsky, D. W., Beckley, A. L., Arseneault, L., Barnes, J. C., … Moffitt, T. E. (2018). Genetics and crime: Integrating new genomic discoveries into psychological research about antisocial behavior. *Psychological Science, 29*(5), 791–803. https://doi.org/10.1177/0956797617744542
White, E. K., Garon-Carrier, G., Tosto, M. G., Malykh, S. B., Li, X., Kiddle, B., … Kovas, Y. (2018). Twin classroom dilemma: To study together or separately? *Developmental Psychology, 54*(7), 1244–1254. https://doi.org/10.1037/dev0000519
Zahn-Waxler, C., Klimes-Dougan, B., & Slattery, M. J. (2000). Internalizing problems of childhood and adolescence: Prospects, pitfalls, and progress in understanding the development of anxiety and depression. *Development and Psychopathology, 12*, 443–466. https://doi.org/10.1017/S0954579400003102

The Impact of School Social Experiences on Socioemotional and Behavioral Problems: The Hypothesized Role of DNA Methylation

Maude Comtois-Cabana, J. Marieke Buil, Nadine Provençal, and Isabelle Ouellet-Morin

Introduction

Beyond the influence of the family environment, research shows that peer relationships contribute, positively and negatively, to the children's trajectories of social, emotional, and behavioral development (Rubin et al., 2006). Adverse peer relationships include peer rejection, peer victimization, and affiliation with deviant peers (see chapters "Prologue: Introduction" and "Elementary School Social Experiences with Peers and Teachers: Manifestation and Development", this volume). In this chapter, we will focus solely on peer victimization, for which the prevalence rate is estimated at 12.6% worldwide (Craig et al., 2009). A child is being victimized when he or she is exposed repeatedly and chronically to hurtful actions perpetrated by a peer or a group of peers between whom there is an imbalance of power (Olweus, 1994). Peer victimization occurs mostly in school settings, where children spend most of their time (Arseneault et al., 2010). Actions include physical aggression (e.g., hitting), verbal aggression (e.g., name-calling), and relational aggression (e.g., social exclusion) (Crick & Grotpeter, 1996). Longitudinal studies indicated that being victimized by peers in childhood may have serious consequences on health

M. Comtois-Cabana
Department of Psychology, University of Montreal, Montreal, QC, Canada

J. M. Buil
Department of Clinical, Neuro and Developmental Psychology, Vrije Universiteit & Amsterdam Public Health Research Institute, Amsterdam, The Netherlands

N. Provençal
Faculty of Health Sciences, Simon Fraser University, Burnaby & BC Children's Hospital Research Institute, Vancouver, BC, Canada

I. Ouellet-Morin (✉)
School of Criminology, University of Montreal & Research Center of the Montreal Mental Health University Institute, Montreal, QC, Canada
e-mail: isabelle.ouellet-morin@umontreal.ca

P. A. C. van Lier, K. Deater-Deckard (eds.), *Biosocial Interplay During Elementary School*, https://doi.org/10.1007/978-3-031-07109-6_5

and functioning throughout the life course, including mental and physical health problems (Arseneault, 2018). A meta-analysis conducted by Reijntjes et al. (2010), comprising the data of 12,361 children, has indeed documented that children who were the target of peer victimization subsequently exhibited higher levels of loneliness, withdrawal behaviors, depression, and anxiety. Higher levels of physical aggression and attention problems have also been noted in a second meta-analysis totalizing the data of 5825 children (Reijntjes et al., 2011).

Longitudinal studies thus support the idea that peer victimization exerts a detrimental impact on several domains of functioning. Importantly, these studies have broadened our understanding of the consequences that may follow peer victimization while clarifying the temporal sequence of events, whereby peer victimization was associated with increased risk of adjustment difficulties later in life. However, genetic factors and other environmental factors could influence the magnitude of these effects. For example, the children's genetic background could indirectly affect their sensitivity to stress or vulnerability to exhibit social, emotional, and behavioral problems when victimized by their peers (see chapter "How Peers and Teachers Shape Elementary School Children's Academic and Socioemotional Development", this volume). To this end, the discordant monozygotic (MZ) twin design is often regarded as a rigorous research design enabling a stronger control for a wider range of confounders. By contrasting genetically identical children—MZ twins who grew up together in the same families, but who have been exposed to distinct environments—researchers can estimate association between a specific environmental exposure (e.g., peer victimization) on functioning (e.g., aggressive behaviors), over and above the children's genetic background and shared environmental influences (Vitaro et al., 2009). Using this design, Silberg et al. (2016) reported, in a sample of 145 MZ twin pairs discordant for peer victimization, that victimized twins had higher levels of anxiety, separation anxiety, as well as attention deficit hyperactivity disorder (ADHD) in childhood and higher rates of suicidal ideation in adulthood compared to their non-victimized co-twins. According to the same research design, Brendgen et al. (2013) showed that the higher levels of depressive symptoms and aggressive behaviors displayed by the victims were independent of (i.e., could not be explained by) their genetic risk for depression and aggression, once more supporting the hypothesized impact of peer victimization on emotional and behavioral problems.

Although numerous studies have suggested that peer victimization may detrimentally affect present and future functioning, as well as the well-being of the victims, little is known about the biological processes underlying these associations. Considering that peer victimization, unlike other adverse experiences that may occur during childhood (e.g., maltreatment), is often perceived by the victims as novel, unpredictable, uncontrollable, or threatening one's physical and social self, researchers have hypothesized that such experiences may affect functioning because they jeopardize stress-related biological systems. Cumulative evidence generally supports this hypothesis. New lines of research are now pushing forward the frontier of knowledge to identify the molecular mechanisms by which peer victimization "gets under the skin and cells."

In this chapter, we first briefly examine how the neuroendocrine system may be affected by peer victimization. Next, we describe the extent to which epigenetic mechanisms, especially DNA methylation, may be altered by life experiences and could, as such, jeopardize later development, a concept referred to as biological embedding (Hertzman, 2012). Finally, we outline key methodological, biological, and statistical limitations confining the generalization of the emerging findings investigating epigenetic mechanisms in peer relationships research.

Peer Victimization: A Stressful Experience?

The hypothalamic-pituitary-adrenocortical (HPA) axis is one of the main systems underlying the physiological response to stress (see chapter "The Impact of School Social Experiences on Socioemotional and Behavioral Problems: The Hypothesized Role of DNA Methylation ", this volume). Both physical and psychological stressors, real or perceived, have thus the potential to induce the release of cortisol, a glucocorticoid hormone secreted by the HPA axis. Acute stress leads to a short-term activation of the HPA axis, resulting in temporary elevations of cortisol (Koss & Gunnar, 2018). Conversely, chronic stress leads to a prolonged exposure of the body and the brain to either enhanced or blunted secretion of cortisol later on, which are both potentially damaging to the organism (Koss & Gunnar, 2018).

Because of its repetitive and nature, peer victimization can be viewed as a form of chronic stress. Subjective reports of children victimized by their peers have indeed indicated that these experiences are perceived as highly stressful (Östberg et al., 2018). Accordingly, it has been hypothesized that peer victimization induces stable disruptions in cortisol secretion in basal and stressful contexts, which could eventually jeopardize other neurophysiological systems involved in emotional and behavioral regulation (Vaillancourt, 2018). Support for this hypothesis can be found in the knowledge that glucocorticoid receptors, to which the glucocorticoid stress hormone cortisol preferably binds, are found in several areas of the brain underlying emotional and behavioral regulation, such as the amygdala, the hippocampus, and the prefrontal cortex (de Kloet et al., 2005). A growing body of evidence suggests that children and adolescents who have been victimized by their peers have lower levels of cortisol secretion during the day (Knack et al., 2011; Östberg et al., 2018; Vaillancourt et al., 2008), in response to stress (Calhoun et al., 2014; Knack et al., 2011; Ouellet-Morin et al., 2011a, b), and lower and higher cortisol secretion over an extended period of time, as measured in hair (Ouellet-Morin et al., 2020). Nonetheless, inconsistent findings are reported, pointing either to higher (Chen et al., 2018) or to nonsignificant differences in cortisol responses between victims and non-victims (Hamilton et al., 2008; Katz et al., 2019; Rudolph et al., 2010, 2011). These findings echo other studies conducted with rodents and humans, which also reported dysregulated patterns of stress hormone secretion in basal and stressful contexts following early adversity, such as low maternal care in rodents (e.g., Liu, 1997) and child maltreatment in humans (e.g., Bernard et al., 2017; Bunea et al., 2017). In addition to understand why lower, and sometimes higher, cortisol secretion is noted in individuals with a history of peer victimization, researchers ought to identify possible molecular mechanisms that bring about these differences, as well as the consequences these "biological traces" may have on socioemotional and behavioral functioning and health over time.

Biological Embedding of Stress Through the Epigenome

In molecular biology, the term "epigenome" refers to the set of epigenetic modifications to the DNA, histone proteins, and chromatin structure (Feil & Fraga, 2012). Unlike genetic mutations, epigenetic modifications do not alter the DNA sequence and are thus potentially reversible. Furthermore, epigenetic modifications can be inherited and transmitted during cell divisions (Radford, 2018). The main function of the epigenome is to regulate the expression of genes, the process by which the DNA sequence of a gene is converted into a protein (Provençal & Binder, 2015). DNA methylation may interfere with gene expression. In other words, while genes provide the instructions to synthesize proteins, epigenetic modifications can influence the cell's ability to read these instructions and to carry them out efficiently.

DNA methylation is currently the most studied epigenetic modification in humans because it is relatively stable over time and easily quantifiable (Jones et al., 2018). The DNA sequence consists of four bases, cytosine (C), guanine (G), adenine (A), and thymine (T). DNA methylation involves the addition of a methyl group (CH_3) to a cytosine base paired with a guanine base (i.e., the reference to CpG sites). DNA methylation regulates the expression of genes in two ways. First, DNA methylation interferes with the binding of transcription factors to the DNA sequence, which are proteins that initiate gene expression (Bird, 2002). Second, DNA methylation attracts proteins that restrain even more gene expression (Bird, 2002). Depending on its location, DNA methylation may have different effects on gene expression. Gene expression can be either "turned on," resulting in increased levels of protein synthesis, or "turned off," resulting in decreased levels of protein synthesis. In general, when CpG sites are methylated in the promoter region of a gene, which is the region where gene transcription is initiated, the expression of that gene is "turned off" (Bird, 1986). However, when CpG sites are methylated in the body of a gene, which is the region that contains the DNA segment to be transcribed, gene expression could be either "turned on" or "turned off" (Jiang et al., 2013; Jjingo et al., 2012).

Researchers have investigated DNA methylation patterns according to two approaches. In early DNA methylation studies, researchers have adopted a candidate gene approach, involving the preselection of genes that are hypothesized to be associated with the variables of interest (e.g., depressive symptoms) or involved in neurobiological systems (e.g., emotion regulation) or in mechanisms of action of drugs (e.g., selective serotonin reuptake inhibitors) thought to affect these outcomes. Recently, epigenome-wide association studies (EWAS) or methylome-wide association studies (MWAS) have gained popularity as they allow researchers to examine the entire epigenome or methylome, thus providing the opportunity to discover novel epigenetic variations related to certain environments or phenotypes of interest.

Although the use of brain tissues is often judged preferable to assess the impact that DNA methylation may have on behaviors, processes, and characteristics mediated by the brain, such as psychological functioning, DNA methylation cannot be measured directly in the brain of living humans. Researchers are thus forced to use peripheral tissues, such as blood or buccal cells, as surrogates for brain tissues in living individuals. The most common technique to measure DNA methylation levels from biological samples is the sodium bisulfite treatment, allowing to quantify DNA methylation as a percentage of unmethylated versus methylated sites (for an extended description of the laboratory techniques used to measure DNA methylation, see Jones et al., 2018).

As illustrated in Fig. 1, DNA methylation patterns are primarily influenced by the genome (i.e., the DNA sequence). Nonetheless, DNA methylation patterns are not fixed. Accordingly, dynamic changes in DNA methylation patterns can occur during development in response to environmental signals emanating from inside (e.g., neurotransmitters, hormones) or outside (e.g., diet, pollutants) the organism, especially in utero and during the first years of life (Meaney, 2010). As displayed in Fig. 1, these sensitive periods of development are characterized by enhanced plasticity to environmental signals because the brain is still immature and undergoes

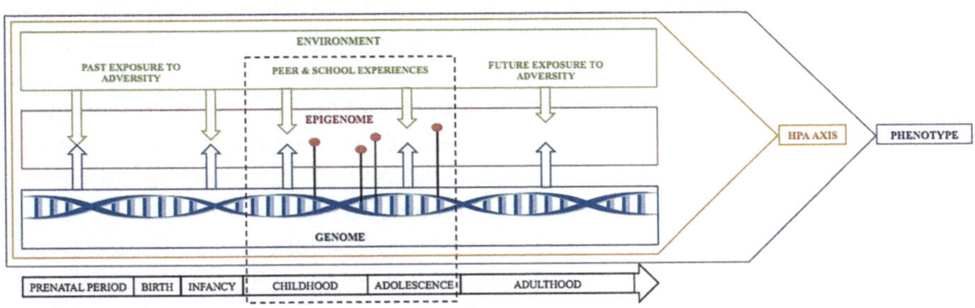

Fig. 1 Conceptual model of the interaction between the genome and the environment across development. The genome (*blue box*) includes all the genetic material (DNA) of an organism, providing all the information essential for functioning. The blue arrows represent the stability of the genome's influence on DNA methylation patterns throughout development. The environment (*green box*) comprises various environmental influences, such as peer victimization. The green arrows illustrate the decline of the environment's influence on DNA methylation patterns across development. Indeed, the epigenome is highly responsive to environmental exposures during sensitive periods of development, such as prenatal and early postnatal periods. The epigenome (*red box*) includes all the chemical modifications to the DNA and histone proteins of an organism. The red dots represent epigenetic modifications, such as DNA methylation, which may influence gene expression, and thus the activity of several neurophysiological stress-related systems, i.e., biological embedding (*orange box*). Together, these four factors contribute to the emergence of vulnerability to stress and individual's propensity to experience socioemotional, behavioral, and health problems over the lifespan (*purple box*)

rapid development and synaptic pruning, a natural process during which the brain eliminates extra synapses (Boyce & Kobor, 2015). In this manner, early social experiences could more readily influence DNA methylation than experiences occurring later in life. Altogether, these experiences could positively or negatively affect several neurophysiological systems supporting adaptation to the environment, including the HPA axis. DNA methylation thus represents a promising mechanism to better understand how peer victimization may have lasting consequences on socioemotional development, behaviors, and health.

Early-Life Stress and DNA Methylation

The influence of social interactions on the epigenome was initially studied in rodents. Notably, it was shown that adult offspring exposed to less maternal care in the first weeks of life showed higher methylation levels at the exon 1_7 of the *NR3C1* gene and a reduced expression of this gene (Weaver et al., 2004). The *NR3C1* gene encodes glucocorticoid receptors (GRs), which help to regulate the activity and the effect of the HPA axis by binding to glucocorticoids such as cortisol (Kino & Chrousos, 2002). Based on the rodent maternal care model, McGowan et al. (2009) tested whether humans exposed to child maltreatment exhibited distinct DNA methylation patterns in hippocampal tissues of adults who committed suicide, as compared to adult suicide victims who did not have a history of child maltreatment. As expected, individuals who were maltreated as children showed higher levels of methylation of the exon 1_F of the *NR3C1* gene, the homolog region of the exon 1_7 in rodents. In contrast, however, the majority of studies conducted in humans have relied on peripheral tissues, such as blood, saliva, and buccal cells. In a systematic review, a majority of these studies (89%) reported similar findings, that is, higher levels of methylation of the exon 1_F within the *NR3C1* gene among individuals exposed to early-life adversity (Turecki & Meaney, 2016).

DNA methylation patterns were also investigated in other stress-related genes. For instance, Beach et al. (2010) found that adults who were physically and sexually abused as children had higher methylation levels within the promoter region of the *SLC6A4* gene. Notably, the *SLC6A4* gene is hypothesized to be involved in impulsivity and aggression behavior and is the target of many antidepressant medications (Coleman & Gouaux, 2018). Importantly, higher levels of methylation of the *SLC6A4*gene have been associated with reduced expression of the serotonin transporter (Philibert et al., 2007), albeit not consistently so (Duman & Canli, 2015). These findings thus partially support the putative impact that changes in DNA methylation may have on later socioemotional and behavioral difficulties following exposure to early adversity.

Emerging Evidence of Associations Between Peer Victimization and DNA Methylation

Building on the previously described evidence drawn from animal models of maternal care and studies conducted with humans in the context of child maltreatment, peer relationships researchers proposed that DNA methylation may also partly explain how peer victimization increases risk of socioemotional and behavioral problems later in life (Vaillancourt, 2018; Vaillancourt et al., 2013). So far, only a handful of studies have investigated the association between peer victimization and DNA methylation. These studies have adopted either a candidate gene or methylome-wide approach.

In a first study, Ouellet-Morin et al. (2013) studied DNA methylation patterns of the *SLC6A4* gene from buccal cells in 28 MZ twin pairs discordant for peer victimization in elementary school. Both groups exhibited similar DNA methylation patterns prior to peer victimization, at 5 years of age. However, compared to their non-victim co-twins, twins who were victimized by their peers in elementary school had, on average, higher levels of DNA methylation at 10 years of age at a particular CpG site within the promoter region of the *SLC6A4* gene. The difference was, however, small between the twins (i.e., 4% differences in methylation level). Nonetheless, the difference in DNA methylation noted between these groups was notable because it could not be attributed to children's genetic makeup or shared family environments due to the discordant monozygotic twin design. Moreover, twins who exhibited higher levels of DNA methylation at this CpG site at 10 years had lower cortisol responses to stress 2 years later, at 12 years of age. This study was the first to provide support to the idea that peer victimization may induce changes in DNA methylation.

Conversely, Mulder et al. (2020) who studied the entire epigenome using peripheral blood collected before and after bullying victimization in two longitudinal population cohorts (totalizing 1352 children) found no evidence for association with methylation levels in the *SLC6A4* or *NR3C1* genes, which contrasted with the previous study, as well as with other reports of associations between early adversity and higher *NR3C1* methylation in rodent and human studies (rodent: Weaver et al., 2004; human: McGowan et al., 2009; Turecki & Meaney, 2016). They, however, found that bullying exposure was associated with a small (between 0.12% and 0.21%) but significant decrease in DNA methylation over time in a CpG site annotated to *RAB14* (e.g., important for cellular signaling) while methylation levels were increased at that CpG site in non-bullied children during the same period. This association remained significant after controlling for a wide range of potential confounders, including the exposure to stressful events other than bullying and alcohol use.

In addition, a population-based study including 2232 children did not either find evidence for associations between peer victimization in childhood and adolescence and methylome-wide profiles of DNA extracted from peripheral blood sampled at 18 years of age, once adequate control for confounders was applied (Marzi et al.,

2018). Furthermore, additional associations were tested for six candidate genes, including the *SLC6A4* gene. Only a few significant associations were detected, including with the *SLC6A4* gene. While these mixed findings suggest that peer victimization may not have a pervasive effect on methylation profile, it should be noted that this inconsistency may also arise from the distinct approaches used in these studies. Indeed, the CpG sites examined in Ouellet-Morin et al.' (2013) relied on a candidate gene approach conducted in childhood (age 10), while Mulder et al. (2020) and Marzi et al. (2018) investigated differential DNA methylation patterns across the epigenome in childhood (6 and 10 years) and early adulthood (age 18), respectively. These differences limit the direct comparison of the findings.

Peer Victimization and DNA Methylation: A Focus on Adjustment Difficulties

While informative, the studies we just described focused only on the association between peer victimization and differences in DNA methylation. We thus don't know whether differences in DNA methylation explain higher levels of social, emotional, and behavioral problems noted in individuals who have been bullied. Efstathopoulos et al. (2018) first investigated these tripartite associations between peer victimization, methylation levels of the *NR3C1* gene, and internalizing symptoms in saliva samples among 1149 adolescents aged between 13 and 14 years old. Peer victimization was associated with higher DNA methylation levels at one of the five CpG sites investigated within the exon 1_F of the *NR3C1* gene, with small but significant mean difference between the victims and non-victims (i.e., 0.37%). Furthermore, higher levels of DNA methylation at this CpG site, as well as two others, were associated with self-reported symptoms of anxiety and depression. Yet still, because peer victimization, DNA methylation levels, and internalizing symptoms were assessed simultaneously, the directionality of these associations remains unclear and the direct tests of the presumed mediation hypothesis were not conducted.

In addition, Buil et al. (submitted for publication) investigated associations between DNA methylation levels at birth, age 7, and ages 15–17 with chronic peer victimization throughout childhood and various forms of psychopathology at ages 7 and 15 in a population sample of 936 children followed up prospectively from birth to adolescence. The study showed that children who were persistently victimized by their peers throughout the elementary school period had higher levels of DNA methylation in both *SLC6A4* and *NR3C1* genes in adolescence as compared to non-victimized children. Notably, these differences were not present at birth. Similarly to the previous studies, the magnitude of these effects was small (i.e., ranging from 0.31% to 0.35% of differences between the victims and non-victims). Furthermore, chronic peer victimization was associated with increasing levels of generalized anxiety from childhood to adolescence, which was partially explained by the increasing levels of *SLC6A4* methylation at ages 15–17. This indirect effect remained over and above a wide range of environmental risk factors (e.g., prenatal stressors, child maltreatment, and the participants' substance use).

In sum, three candidate gene studies suggested that peer victimization is associated with changes in methylation of stress-related genes and that such changes may signal differences in stress reactivity or in the presence of emotional difficulties. Importantly, however, these findings have not been replicated in the epigenome-wide association studies and should, as such, be considered with caution.

Methodological, Biological, and Statistical Considerations

In this section, we present several challenges inherent to epigenetic studies and we propose recommendations to improve our understanding of the role of DNA methylation in the onset (or exacerbation) of adjustment difficulties following peer victimization.

Methodological Considerations Important potential confounds are often overlooked in DNA methylation studies and could partly explain inconsistent findings, including genetic factors. Teh et al. (2014) and Czamara et al. (2019) showed that the majority of DNA methylation variation arises as a result of an interaction between genetic and environmental factors. In other words, the association between peer victimization and DNA methylation may vary according to the children's genetic background. Future studies should thus consider genetic factors to ascertain with greater accuracy the role of DNA methylation in socioemotional and behavioral functioning following peer victimization. Second, studies should adjust for covariates, such as age, sex, the use of medication, and the use of alcohol, tobacco, or drugs (Jones et al., 2018). This echoes findings reported by Fraga et al. (2005) who showed that older MZ twin pairs exhibited larger differences in DNA methylation patterns in comparison with younger MZ twins, which was argued to arise randomly with time and as a function of exposure to distinct environments as the twins grow apart. Furthermore, Yousefi et al. (2015) found that newborn boys and girls exhibited distinct DNA methylation patterns at 3% of the 450,000 CpG sites analyzed, pointing to sex differences in methylation. A stringent control for a variety of potential confounders is thus warranted to properly estimate the magnitude of the association between peer victimization, DNA methylation, and socioemotional and behavioral problems.

Biological Considerations Inconsistency in the previous findings may also be related to tissue specificity, because different types of tissues (e.g., whole saliva, blood, buccal epithelial cells) show distinct patterns of methylation. In fact, the type of tissues best predicts differences in DNA methylation patterns between and within individuals (Farré et al., 2015). While the brain is proposed to have effects on social, emotional, and behavioral functioning following peer victimization, DNA methylation cannot be measured in the living human brain for obvious reasons. Researchers must therefore use peripheral tissues, such as blood, saliva, and buccal cells, as surrogates for brain tissues in living individuals. Consequently, it is unclear whether the reported differences (or absence of differences) in DNA methylation patterns in peripheral tissues represent actual differences between the victims and non-victims in the brain (Jones et al., 2018). Smith et al. (2015) reported that DNA methylation patterns are more similar between the DNA extracted from whole saliva and several brain samples (i.e., cerebellum, frontal cortex, entorhinal cortex, and superior temporal gyrus), in comparison with those noted between blood or brain samples. The use of distinct types of tissues may thus, in theory, underline part of the inconsistent findings. Replication attempts should thus target DNA collected from the same tissues or be investigated systematically across several tissues.

Statistical Considerations Existing studies suggest that peer victimization (but also child maltreatment) is associated with relatively small differences (or changes) in DNA methylation patterns, with differences ranging from less than 1% to 10% between exposed and unexposed participants (Breton et al., 2017). Therefore, researchers must design studies with adequate statistical power to be able to detect such small differences in DNA methylation, if they exist, particularly when studying epigenome-wide associations implying a larger number of tests (i.e., more than 450,000 sites tested; Jones et al., 2018). Furthermore, researchers should ascertain whether the small differences (or changes) noted in DNA methylation are biologically meaningful and yield distinct profiles of gene expression, for instance.

Altogether, these methodological, biological, and statistical considerations represent important challenges currently limiting the investigation of the presumed associations between peer victimization (or, more generally, social adversity), DNA methylation, and later difficulties. Nevertheless, they also represent promising avenues to improve the quality of the research conducted thus far to shed some light on the molecular mechanisms underlying the long-term effects peer victimization have on social, emotional and behavioral, and health problems.

Conclusion

Preliminary evidence suggests that peer victimization may be associated with distinct patterns of DNA methylation in stress-related candidate genes, although inconsistent findings have been reported and many limitations constrain these findings. Changes in DNA methylation patterns following early adverse experiences may bear long-lasting consequences on the stress-related biological systems, as well as brain neuronal development, activity, and connectivity (Heim & Binder, 2012). Moreover, only a few studies have formally tested whether individual differences in DNA methylation explain, at least partially, the onset or increasing levels of social, emotional, and behavioral problems following peer victimization. Future research should also consider a wider range of potential confounders, including differences present at the DNA levels, to capture more precisely the magnitude of the associations. Replication should be prioritized, along with the improvement of the measurement of peer victimization. We should also try to measure gene expression and protein synthesis, in addition to DNA methylation, to refine our understanding of the biological pathways involved in these associations and to determine whether the small differences reported in DNA methylation between victims and non-victims are biologically meaningful.

Notwithstanding these methodological, biological, and statistical challenges, the hypothesis that DNA methylation is a molecular pathway by which peer victimization increases vulnerability to stress and risk for social, emotional, and behavioral difficulties later in life still represents an avenue of research for which only time will tell whether it has hold a piece of the puzzle to better understand how peer victimization may jeopardize later well-being.

References

Arseneault, L. (2018). Annual research review: The persistent and pervasive impact of being bullied in childhood and adolescence: Implications for policy and practice. *Journal of Child Psychology and Psychiatry, 59*(4), 405–421.

Arseneault, L., Bowes, L., & Shakoor, S. (2010). Bullying victimization in youths and mental health problems: 'Much ado about nothing'? *Psychological Medicine, 40*(5), 717–729.

Beach, S. R. H., Brody, G. H., Todorov, A. A., Gunter, T. D., & Philibert, R. A. (2010). Methylation at SLC6A4 is linked to family history of child abuse: An examination of the Iowa adoptee sample. *Journal of Medical Genetics, 1*(2), 710–713.

Bernard, K., Frost, A., Bennett, C. B., & Lindhiem, O. (2017). Maltreatment and diurnal cortisol regulation: A meta-analysis. *Psychoneuroendocrinology, 78*, 57–67.

Bird, A. P. (1986). CpG-rich islands and the function of DNA methylation. *Nature, 321*(6067), 209–213.

Bird, A. (2002). DNA methylation patterns and epigenetic memory. *Genes & Development, 16*(1), 6–21.

Boyce, W. T., & Kobor, M. S. (2015). Development and the epigenome: The 'synapse' of gene-environment interplay. *Developmental Science, 18*(1), 1–23.

Brendgen, M., Vitaro, F., Barker, E. D., Girard, A., Dionne, G., Tremblay, R. E., & Boivin, M. (2013). Do other people's plights matter? A genetically informed twin study of the role of social context in the link between peer victimization and children's aggression and depression symptoms. *Developmental Psychology, 49*(2), 327–340.

Breton, C. V., Marsit, C. J., Faustman, E., Nadeau, K., Goodrich, J. M., Dolinoy, D. C., et al. (2017). Small-magnitude effect sizes in epigenetic end points are important in children's environmental health studies: The Children's environmental health and disease prevention research center's Epigenetics Working Group. *Environmental Health Perspectives, 125*(4), 511–526.

Buil, J. M., Cecil, C. M., van Lier, P. A. C., Relton, C. L., & Barker, E. D. (unpublished findings). Relational peer-victimization, DNA methylation and mental health problems: A longitudinal study spanning birth to adolescence.

Bunea, I. M., Szentágotai-Tătar, A., & Miu, A. C. (2017). Early-life adversity and cortisol response to social stress: A meta-analysis. *Translational Psychiatry, 7*(12), 1274.

Calhoun, C. D., Helms, S. W., Heilbron, N., Rudolph, K. D., Hastings, P. D., & Prinstein, M. J. (2014). Relational victimization, friendship, and adolescents' hypothalamic–pituitary–

adrenal axis responses to an in vivo social stressor. *Development and Psychopathology, 26*(3), 605–618.
Chen, G., Kong, Y., Deater-Deckard, K., & Zhang, W. (2018). Bullying victimization heightens cortisol response to psychosocial stress in Chinese children. *Journal of Abnormal Child Psychology, 46*(5), 1051–1059.
Coleman, J. A., & Gouaux, E. (2018). Structural basis for recognition of diverse antidepressants by the human serotonin transporter. *Nature Structural & Molecular Biology, 25*(2), 170–175.
Craig, W., Harel-Fisch, Y., Fogel-Grinvald, H., Dostaler, S., Hetland, J., Simons-Morton, B., et al. (2009). A cross-national profile of bullying and victimization among adolescents in 40 countries. *International Journal of Public Health, 54*(Suppl 2), 216–224.
Crick, N. R., & Grotpeter, J. K. (1996). Children's treatment by peers: Victims of relational and overt aggression. *Development and Psychopathology, 8*(2), 367–380.
Czamara, D., Eraslan, G., Page, C. M., Lahti, J., Lahti-Pulkkinen, M., Hämäläinen, E., et al. (2019). Integrated analysis of environmental and genetic influences on cord blood DNA methylation in new-borns. *Nature Communications, 10*(1), 1–18.
de Kloet, E. R., Sibug, R. M., Helmerhorst, F. M., & Schmidt, M. (2005). Stress, genes and the mechanism of programming the brain for later life. *Neuroscience and Biobehavioral Reviews, 29*, 271–281.
Duman, E. A., & Canli, T. (2015). Influence of life stress, 5-HTTLPR genotype, and SLC6A4 methylation on gene expression and stress response in healthy Caucasian males. *Biology of Mood & Anxiety Disorders, 5*(1), 2.
Efstathopoulos, P., Andersson, F., Melas, P. A., Yang, L. L., Villaescusa, J. C., Rüegg, J., et al. (2018). NR3C1 hypermethylation in depressed and bullied adolescents. *Translational Psychiatry, 8*(1), 121.
Farré, P., Jones, M. J., Meaney, M. J., Emberly, E., Turecki, G., & Kobor, M. S. (2015). Concordant and discordant DNA methylation signatures of aging in human blood and brain. *Epigenetics & Chromatin, 8*(1), 19.
Feil, R., & Fraga, M. F. (2012). Epigenetics and the environment: Emerging patterns and implications. *Nature Reviews Genetics, 13*(2), 97–109.
Fraga, M. F., Ballestar, E., Paz, M. F., Ropero, S., Setien, F., Ballestar, M. L., et al. (2005). Epigenetic differences arise during the lifetime of monozygotic twins. *Proceedings of the National Academy of Sciences of the United States of America, 102*(30), 10604–10609.
Hamilton, L. D., Newman, M. L., Delville, C. L., & Delville, Y. (2008). Physiological stress response of young adults exposed to bullying during adolescence. *Physiology & Behavior, 95*(5), 617–624.
Heim, C., & Binder, E. B. (2012). Current research trends in early life stress and depression: Review of human studies on sensitive periods, gene–environment interactions, and epigenetics. *Experimental Neurology, 233*(1), 102–111.
Hertzman, C. (2012). Putting the concept of biological embedding in historical perspective. *Proceedings of the National Academy of Sciences, 109*(Supplement_2), 17160–17167.
Jiang, Y., Liu, S., Chen, X., Cao, Y., & Tao, Y. (2013). Genome-wide distribution of DNA methylation and DNA demethylation and related chromatin regulators in cancer. *Biochimica et Biophysica Acta (BBA) – Reviews on Cancer, 1835*(2), 155–163.
Jjingo, D., Conley, A. B., Yi, S. V., Lunyak, V. V., & Jordan, I. K. (2012). On the presence and role of human gene-body DNA methylation. *Oncotarget, 3*(4), 462–474.
Jones, M. J., Moore, S. R., & Kobor, M. S. (2018). Principles and challenges of applying epigenetic epidemiology to psychology. *Annual Review of Psychology, 69*(1), 459–485.
Katz, D. A., Peckins, M. K., & Lyon, C. C. (2019). Adolescent stress reactivity: Examining physiological, psychological and peer relationship measures with a group stress protocol in a school setting. *Journal of Adolescence, 74*, 45–62.
Kino, T., & Chrousos, G. P. (2002). Tissue-specific glucocorticoid resistance-hypersensitivity syndromes: Multifactorial states of clinical importance. *Journal of Allergy and Clinical Immunology, 109*(4), 609–613.

Knack, J. M., Jensen-Campbell, L. A., & Baum, A. (2011). Worse than sticks and stones? Bullying is associated with altered HPA axis functioning and poorer health. *Brain and Cognition, 77*(2), 183–190.

Koss, K. J., & Gunnar, M. R. (2018). Annual research review: Early adversity, the hypothalamic–pituitary–adrenocortical axis, and child psychopathology. *Journal of Child Psychology and Psychiatry, 59*(4), 327–346.

Liu, D. (1997). Maternal care, hippocampal glucocorticoid receptors, and hypothalamic-pituitary-adrenal responses to stress. *Science, 277*(5332), 1659–1662.

Marzi, S. J., Sugden, K., Arseneault, L., Belsky, D. W., Burrage, J., Corcoran, D. L., et al. (2018). Analysis of DNA methylation in young people: Limited evidence for an association between victimization stress and epigenetic variation in blood. *American Journal of Psychiatry, 175*(6), 517–529.

McGowan, P. O., Sasaki, A., D'alessio, A. C., Dymov, S., Labonté, B., Szyf, M., et al. (2009). Epigenetic regulation of the glucocorticoid receptor in human brain associates with childhood abuse. *Nature Neuroscience, 12*(3), 342.

Meaney, M. J. (2010). Epigenetics and the biological definition of gene × environment interactions. *Child Development, 81*(1), 41–79.

Mulder, R. H., Walton, E., Neumann, A., Houtepen, L. C., Felix, J. F., Bakermans-Kranenburg, M. J., et al. (2020). Epigenomics of being bullied: Changes in DNA methylation following bullying exposure. *Epigenetics, 15*(6-7), 750–764.

Olweus, D. (1994). Bullying at school. In L. R. Huesmann (Ed.), *Aggressive behavior: Current perspectives* (pp. 97–130). Springer US.

Östberg, V., Låftman, S. B., Modin, B., & Lindfors, P. (2018). Bullying as a stressor in mid-adolescent girls and boys–associations with perceived stress, recurrent pain, and salivary cortisol. *International Journal of Environmental Research and Public Health, 15*(2), 364.

Ouellet-Morin, I., Danese, A., Bowes, L., Shakoor, S., Ambler, A., Pariante, C. M., et al. (2011a). A discordant monozygotic twin design shows blunted cortisol reactivity among bullied children. *Journal of the American Academy of Child & Adolescent Psychiatry, 50*(6), 574–582.e573.

Ouellet-Morin, I., Odgers, C. L., Danese, A., Bowes, L., Shakoor, S., Papadopoulos, A. S., et al. (2011b). Blunted cortisol responses to stress signal social and behavioral problems among maltreated/bullied 12-year-old children. *Biological Psychiatry, 70*(11), 1016–1023.

Ouellet-Morin, I., Wong, C., Danese, A., Pariante, C., Papadopoulos, A., Mill, J., & Arseneault, L. (2013). Increased serotonin transporter gene (SERT) DNA methylation is associated with bullying victimization and blunted cortisol response to stress in childhood: A longitudinal study of discordant monozygotic twins. *Psychological Medicine, 43*(9), 1813–1823.

Ouellet-Morin, I., Cantave, C., Paquin, S., Geoffroy, M.-C., Brendgen, M., Vitaro, F., et al. (2020). Associations between developmental trajectories of peer victimization, hair cortisol and depressive symptoms: A longitudinal study. *Journal of Child Psychology and Psychiatry, 62*(1), 19–27.

Philibert, R., Madan, A., Andersen, A., Cadoret, R., Packer, H., & Sandhu, H. (2007). Serotonin transporter mRNA levels are associated with the methylation of an upstream CpG island. *American Journal of Medical Genetics Part B: Neuropsychiatric Genetics, 144B*(1), 101–105.

Provençal, N., & Binder, E. B. (2015). The effects of early life stress on the epigenome: From the womb to adulthood and even before. *Experimental Neurology, 268*, 10–20.

Radford, E. J. (2018). Exploring the extent and scope of epigenetic inheritance. *Nature Reviews Endocrinology, 14*(6), 345.

Reijntjes, A., Kamphuis, J. H., Prinzie, P., & Telch, M. J. (2010). Peer victimization and internalizing problems in children: A meta-analysis of longitudinal studies. *Child Abuse & Neglect, 34*(4), 244–252.

Reijntjes, A., Kamphuis, J. H., Prinzie, P., Boelen, P. A., van der Schoot, M., & Telch, M. J. (2011). Prospective linkages between peer victimization and externalizing problems in children: A meta-analysis. *Aggressive Behavior, 37*(3), 215–222.

Rubin, K. H., Bukowski, W. M., & Parker, J. G. (2006). Peer interactions, relationships, and groups. In *Handbook of child psychology: Social, emotional, and personality development* (Vol. Vol. 3, 6th ed., pp. 571–645). Wiley.

Rudolph, K. D., Troop-Gordon, W., & Granger, D. A. (2010). Peer victimization and aggression: Moderation by individual differences in salivary Cortiol and alpha-amylase. *Journal of Abnormal Child Psychology, 38*(6), 843–856.

Rudolph, K. D., Troop-Gordon, W., & Granger, D. A. (2011). Individual differences in biological stress responses moderate the contribution of early peer victimization to subsequent depressive symptoms. *Psychopharmacology, 214*(1), 209–219.

Silberg, J. L., Copeland, W., Linker, J., Moore, A. A., Roberson-Nay, R., & York, T. P. (2016). Psychiatric outcomes of bullying victimization: A study of discordant monozygotic twins. *Psychological Medicine, 46*(9), 1875–1883.

Smith, A. K., Kilaru, V., Klengel, T., Mercer, K. B., Bradley, B., Conneely, K. N., et al. (2015). DNA extracted from saliva for methylation studies of psychiatric traits: Evidence tissue specificity and relatedness to brain. *American Journal of Medical Genetics. Part B, Neuropsychiatric Genetics: The Official Publication of the International Society of Psychiatric Genetics, 168B*(1), 36–44.

Teh, A. L., Pan, H., Chen, L., Ong, M.-L., Dogra, S., Wong, J., et al. (2014). The effect of genotype and in utero environment on interindividual variation in neonate DNA methylomes. *Genome Research, 24*(7), 1064–1074.

Turecki, G., & Meaney, M. J. (2016). Effects of the social environment and stress on glucocorticoid receptor gene methylation: A systematic review. *Biological Psychiatry, 79*(2), 87–96.

Vaillancourt, T. (2018). Introduction to the special issue: The neurobiology of peer victimization. *Merrill-Palmer Quarterly, 64*(1), 1–11.

Vaillancourt, T., Duku, E., Decatanzaro, D., Macmillan, H., Muir, C., & Schmidt, L. A. (2008). Variation in hypothalamic–pituitary–adrenal axis activity among bullied and non-bullied children. *Aggressive Behavior: Official Journal of the International Society for Research on Aggression, 34*(3), 294–305.

Vaillancourt, T., Hymel, S., & McDougall, P. (2013). The biological underpinnings of peer victimization: Understanding why and how the effects of bullying can last a lifetime. *Theory Into Practice, 52*(4), 241–248.

Vitaro, F., Brendgen, M., & Arseneault, L. (2009). The discordant MZ-twin method: One step closer to the holy grail of causality. *International Journal of Behavioral Development, 33*(4), 376–382.

Weaver, I. C. G., Cervoni, N., Champagne, F. A., D'Alessio, A. C., Sharma, S., Seckl, J. R., et al. (2004). Epigenetic programming by maternal behavior. *Nature Neuroscience, 7*(8), 847–854.

Yousefi, P., Huen, K., Davé, V., Barcellos, L., Eskenazi, B., & Holland, N. (2015). Sex differences in DNA methylation assessed by 450 K BeadChip in newborns. *BMC Genomics, 16*(1), 911.

Biological Embedding of Peer Experiences: The Contribution of Peer Adversity to Stress Regulation

Karen D. Rudolph, Haley V. Skymba, Haina H. Modi, Megan M. Davis, and Wing Yan Sze

Peer relationships serve as a critical context of development by supporting youths' fundamental need for relatedness. Although healthy peer relationships can operate as a positive socialization force that enhances development, unhealthy peer relationships can operate as a negative socialization force that undermines development. Youth who are exposed to adversity or marginalized from peer groups not only lose the benefits of having a strong social network but also suffer the risks associated with stress exposure (Rudolph et al., 2016a). This chapter focuses on understanding how exposure to peer adversity during the school years upsets normative developmental trajectories and "gets under the skin" to shape biological stress-response systems in ways that forecast both proximal and chronic risk for maladjustment. We also consider the possibility, however, that tuning of biological stress-response systems by early adversity may have the potential to foster better adaptation in the context of later positive peer contexts.

Components of Biological Stress-Response Systems

Biological stress-response systems are viewed as one primary pathway through which experience is translated into long-term developmental adaptations (Ellis et al., 2017). The hypothalamic-pituitary-adrenal (HPA) axis and the autonomic (parasympathetic and sympathetic) nervous system (ANS) mobilize the body to respond in a flexible and adaptive manner to physical and social stressors. In addition, recent research implicates the immune system as an integral component of stress responses within a social context.

K. D. Rudolph (✉) · H. V. Skymba · H. H. Modi · M. M. Davis · W. Y. Sze
University of Illinois, Urbana-Champaign, Champaign, IL, USA
e-mail: krudolph@illinois.edu

P. A. C. van Lier, K. Deater-Deckard (eds.), *Biosocial Interplay During Elementary School*, https://doi.org/10.1007/978-3-031-07109-6_6

Hypothalamic-Pituitary-Adrenal (HPA) Axis

HPA axis function involves both a typical diurnal rhythm and stress-reactive activation. The primary end product of the HPA axis is glucocorticoids (cortisol in humans). Cortisol shows a diurnal rhythm that peaks soon after waking (the cortisol awakening response; CAR) and gradually declines throughout the day (diurnal slope). Both physical and psychological stressors can trigger a complex set of interactions involving release of corticotropin-releasing hormone (CRH) by the hypothalamus, which activates release of adrenocorticotropin hormone (ACTH) by the pituitary gland and culminates in release of cortisol by the adrenal glands. Circulating cortisol exerts negative feedback on the system, allowing for a return to homeostasis (Lovallo & Buchanan, 2017). This stress response typically is measured in terms of the difference between post-stressor and pre-stressor cortisol levels or cortisol output across a stressor (area under the curve; AUC). Although moderate elevation of cortisol in response to acute stress is adaptive in the short term as it mobilizes coping resources, excessive or prolonged activation can threaten mental and physical health, forecasting depressive symptoms as well as diseases such as diabetes and hypertension (Miller et al., 2007).

Autonomic Nervous System (ANS)

The autonomic nervous system is composed of the sympathetic nervous system (SNS), a fast-acting system responsible for the body's increase in physiological arousal and "fight-or-flight" response, and the parasympathetic nervous system (PNS), a regulatory system that modulates the SNS, restores the body to homeostasis, and controls the "rest and digest" functions (Beauchaine, 2015; Berntson et al., 2007; Porges, 2007). The SNS and PNS serve largely reciprocal roles, such that increasing activity in one branch leads to decreasing activity in the other. However, coupled responses also may represent either concurrent increases (i.e., coactivation) or decreases (i.e., coinhibition; Berntson et al., 1991), and these systems can display uncoupled activity. The ANS plays a large role in activating and regulating the cardiovascular (CV) system; heart rate increases reflect both SNS activation and PNS withdrawal, whereas heart rate variability specifically reflects PNS activation (Cacioppo et al., 2007). Both systolic and diastolic blood pressure also can be influenced by either SNS or PNS function (Berntson et al., 2007). Respiratory sinus arrhythmia (RSA)—high-frequency heart rate variability associated with respiratory system activity—reflects PNS activation (Beauchaine, 2015; Porges, 2007). The SNS also plays an activating role in electrodermal activity and skin conductance levels. Finally, salivary biomarkers (e.g., salivary alpha-amylase; sAA) are used as an indicator of SNS activity, specifically. Under conditions of short-term acute threat or stress, activation of the SNS and PNS is largely adaptive by allowing individuals to increase arousal and orient attention, which facilitates engagement with the environment, and then to subsequently reestablish equilibrium (Porges, 2007). However, in the face of chronic stress exposure, severe or prolonged activation or less effective regulation within these systems may occur, leading to adverse developmental outcomes such as poor mental health and physical disease (McLaughlin et al., 2015).

Immune System

The immune system organizes and directs the body's response to physical injury or infection, which initiates a cascade of reactions that lead to inflammatory activity. The inflammatory response is regulated by processes in the brain that signal the immune system even before injury occurs (Slavich & Irwin, 2014). Potential threats of injury, such as danger in the environment, can lead immune response genes to initiate a preparatory response, allowing the mobilization of immune cells that are critical for recovery. Broadly, the immune system includes two branches: the innate system and the adaptive system (Slavich & Irwin, 2014). The innate system, composed of immune cells that consistently circulate throughout the body, is the body's first line of defense. If a pathogen is detected, these cells can respond within a few minutes by initiating a chain of inflammatory processes to promote healing (Barton, 2008). The adaptive system largely responds to threats the innate system cannot manage on its own and acts as a second line of defense to release an abundance of white blood cells (lymphocytes). Because the innate system is the body's primary defense, inflammatory activity occurs not only at sites of damage but also systemically, making this system of interest for understanding immune responses to both physical and social threats in daily life.

When the innate immune system is activated, specific families of receptor cells trigger expression of pro-inflammatory immune response genes, leading to the production of cytokines, a primary outcome that is central to the immune system and the inflammatory response (Slavich & Irwin, 2014). Cytokines coordinate cell communication and alter neurochemical and neuroendocrine processes. Pro-inflammatory cytokines increase, or upregulate, inflammation while anti-inflammatory cytokines decrease, or downregulate, inflammation. Pro-inflammatory cytokines (e.g., tumor necrosis factor-α (TNF-α), interleukin-1 (IL-1), interleukin-6 (IL-6)) stimulate inflammation and serve protective roles by allowing immune cells to eliminate pathogens. Although cytokines can have local and specific effects, they also can lead to systemic responses such as the production of C-reactive protein (CRP), a biomarker of inflammation. Ultimately, inflammatory responses help accelerate healing and limit the spread of infection (Slavich & Irwin, 2014). Although this response is largely adaptive in the short term, if it is continually activated by a perceived or actual prolonged threat, it can exert biological costs. Indeed, such extended activation of the inflammatory response is linked to risk for inflammation-related conditions such as asthma and CV disease (Slavich & Irwin, 2014).

Interrelations Among Stress-Response Systems

Despite their distinct roles, the three components of the stress-response systems jointly guide the body's response to stress. Chronic activation of the innate immune system prompts an upregulation of pro-inflammatory immune response genes and a downregulation of antiviral immune response genes. The experience of stress is converted into an inflammatory response involving both the SNS and HPA axis. The SNS regulates the production of pro-inflammatory cytokines, which leads to an increase in systemic inflammatory activity, whereas the HPA axis suppresses both pro-inflammatory and antiviral immune response genes (Irwin & Cole, 2011). Thus, the SNS and HPA axes work together to produce a reciprocal response. The PNS also modulates immune responses regionally, ultimately preventing excessive inflammation. Collectively, these systems coordinate responses to stress to keep individuals physically safe and biologically healthy.

Theoretical Frameworks of Peer Adversity Effects

Despite a growing body of research examining the impact of peer adversity on developing stress-response systems, the field lacks an integrative theoretical framework that considers both the general effects of early adversity on development and factors that might specifically account for the consequences of exposure to peer adversity. Here, we unite prominent developmental models regarding the effect of early adversity on stress-response systems with social-psychological theories regarding how threats to the human need to belong may compromise development. As we describe next, the emerging evidence suggests that (a) prolonged exposure to stress may shape future responses of the stress-response systems that are downregulated or upregulated and (b) there are stable individual differences (e.g., due to predisposing factors and timing and duration of adversity) that must be considered.

General Developmental Frameworks of Early Adversity Effects

Several contemporary theoretical frameworks posit that emerging biological stress-response systems mediate the long-term impact of early experience on developmental adaptation and maladaptation. Experience-adaptive or developmental programming frameworks suggest that biological systems adapt to environmental input, particularly during sensitive periods of development; this malleability is believed to be developmentally constrained such that systems have difficulty readjusting even in the face of subsequent environmental changes (O'Connor, 2003). For example, the allostatic load model (Juster et al., 2010; similar to the toxic stress model; Shonkoff & Bales, 2011) holds that chronic exposure to adversity results in long-term and potentially permanent "wear and tear" to biological stress-response systems, leading to heightened or blunted stress responsivity that promotes risk for a wide range of health-related difficulties.

A cumulative effects framework proposes that early adversity can exert long-term effects if these effects are reinforced by subsequent events (O'Connor, 2003). Specifically, the impact of later stressors may add to the effects of early adversity (additive effects variant; Fig. 1a) or may be dependent on a history of earlier

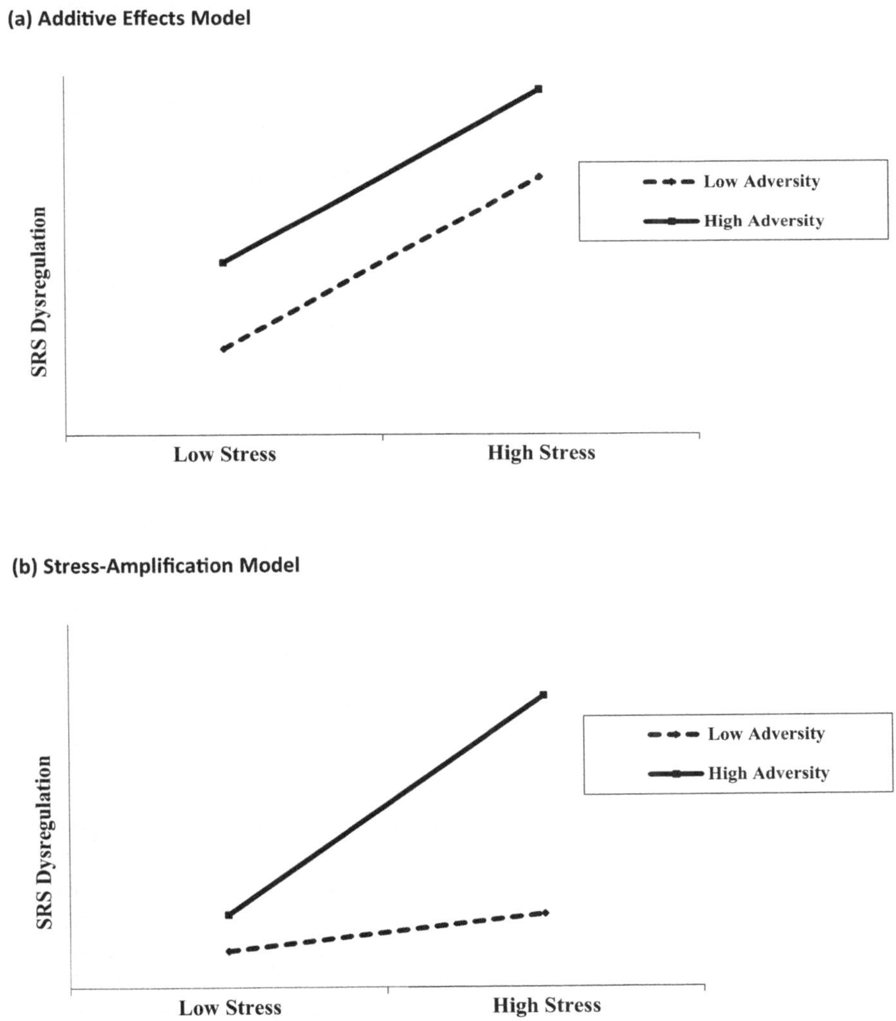

Fig. 1 Conceptual models of the influence of peer adversity on stress-response systems. (**a**) Additive effects model. (**b**) Stress-amplification model. (**c**) Stress-sensitization model. (**d**) Stress-inoculation model

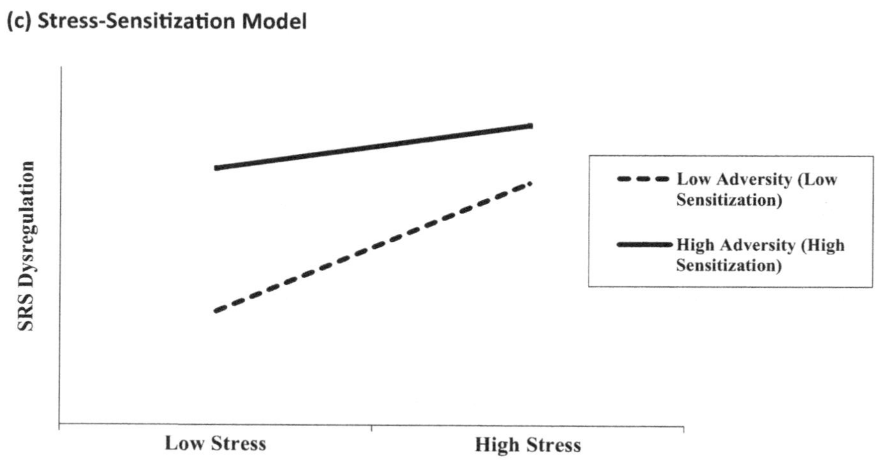

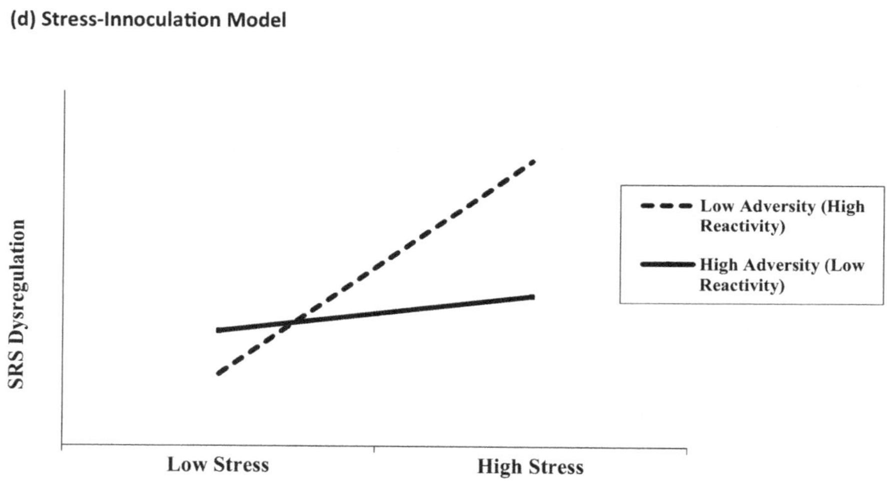

Fig. 6.1 (continued)

adversity (interactive effects variant). Interactive effects models can take several forms (Rudolph et al., 2016a). According to a stress-amplification framework, exposure to early adversity magnifies later stress reactivity, such that youth with a history of early adversity would be more likely to show adverse reactions to subsequent moderate or severe stressors than those without a history of early adversity (Fig. 1b). According to a stress-sensitization framework (Monroe & Harkness, 2005), exposure to early adversity lowers youths' threshold for responding to later stressors, such that youth with a history of early adversity would require milder levels of stress to trigger adverse reactions than those without a history of early adversity (Fig. 1c). According to a stress-inoculation framework, exposure to early adversity steels or buffers youth against later stress, such that youth with a history of early adversity would show less adverse reactions to later stress (i.e., stress resistance) than those without a history of early adversity (Fig. 1d).

Moving beyond a focus on univalent risk or resilience models, contemporary evolutionary biology perspectives consider the functional value of stress-response systems (Ellis et al., 2011, 2017). The biological sensitivity to context theory (Boyce & Ellis, 2005) proposes that early experience (in combination with genetics) shapes the development of individual differences in stress reactivity profiles characterized by either heightened biological sensitivity to context (high plasticity and sensitive stress-response systems) or dampened biological sensitivity to context (low plasticity and unresponsive stress-response systems). Specifically, exposure to either acutely stressful or especially supportive childhood environments is thought to upregulate biological sensitivity to context by enabling children to detect and respond to environmental dangers and threats (within stressful environments) or to detect and respond to environmental resources and support (within favorable environments). Thus, upregulation of stress-response systems would occur in more extreme (adverse or favorable) environments, whereas downregulation of stress-response systems would occur in moderate environments that are neither especially risky nor supportive. In turn, biological sensitivity influences health and behavior in a context-contingent manner, heightening susceptibility to both the health-eroding effects of adverse, threatening contexts and the health-promoting effects of supportive, nurturing contexts.

Similarly, the adaptive calibration model (Ellis et al., 2017) suggests that social experiences guide the development of stress physiology toward adaptive patterns that are conditional upon the context in which an individual develops. Stress-response systems are viewed as information filters that encode environmental cues in ways that amplify (for a highly responsive system) or filter (for a highly unresponsive system) various types of contextual information, both of which can have adaptational costs and benefits. Thus, this model suggests there is no optimal pattern of stress response but rather the adaptiveness of stress-response profiles is context-specific, and individual variation evolves from the contexts in which children develop (Ellis et al., 2017).

Moving beyond the biological sensitivity to context theory to distinguish subtypes of more and less biologically reactive individuals, the adaptive calibration model specifies prototypical patterns of stress reactivity and associated behaviors that may develop in response to particular social environments: (1) sensitive, which develops in safe, predictable, and supportive environments and involves moderate to high stress-response system responsivity, allowing for flexible attention and sensitivity to social feedback; it is predicted to foster positive engagement with the environment and strong self-regulation; (2) buffered, which develops in moderate stressfully environments and involves moderate to low stress-response system responsivity; it is predicted to foster lower anxiety, depression, and risk-taking; (3) vigilant, which develops in stressful contexts (e.g., those with high levels of danger and unpredictability) and involves high levels of stress-response system responsivity, promoting attention to threat; it is predicted to foster either high levels of aggressive and externalizing behaviors (vigilant agonistic pattern; expected more often in males) or high levels of internalizing and withdrawn behaviors (vigilant withdrawn pattern; expected more often in females); and (4) unemotional, which develops in severely stressful or traumatic environments (e.g., those with extreme levels of danger) and involves extreme stress-response systems hyporesponsivity, promoting insensitivity to threat, dangers, and social feedback; it is predicted to foster low empathy, impulsivity, risk-taking, and antisocial behavior, especially in males, and aloof relationships, especially in females. In addition to specifying subtypes of stress responsivity patterns and associated behaviors, the adaptive calibration model builds on biological sensitivity to context theory in several ways by (a) considering the development and functions of basal (tonic) as well as stress-reactive activity of SRS, (b) identifying developmental junctures or "switch points" (i.e., prenatal and early postnatal periods, transition from early to middle childhood, puberty) at which adaptive calibration mechanisms are particularly likely to become activated, and (c) considering gender differences in the origins and meaning of stress responsivity.

Summary Collectively, these models suggest that early adversity can calibrate biological stress-response systems in ways that alter reactivity to later stress. As a result, responses to stress can become either upregulated, indexed by biological sensitization and hyperarousal, or downregulated, indexed by biological desensitization and hypoarousal. The specific profiles of dysregulation may differ according to characteristics of stressors (e.g., severe vs. moderate, chronic vs. acute), characteristics of individuals (e.g., genetic liability, such as a tendency toward under- vs. overregulation of emotions; temperament, such as a tendency toward approach vs. avoidance), stage of development (e.g., transition through "switch points"), and aspects of stress-response systems (e.g., basal activation vs. stress reactivity). Importantly, both biological sensitivity to context theory and the adaptive calibration model propose that individual differences in biological stress-response system responsivity can have trade-offs for health, depending on the conditions under which these differences arose (leading to different subtypes of heightened or dampened reactivity) and the current context in which children are embedded. This point is important to keep in mind when considering how peer adversity-induced changes in stress-response systems may influence future adaptation and maladaptation across childhood.

Specific Frameworks of Peer Adversity Effects

Developmental theories and research regarding the impact of early adversity on stress-response systems mainly focus on challenging experiences within the family (e.g., socioeconomic disadvantage, maladaptive parenting, maltreatment). However, these ideas can be extended to understand pathways through which peer adversity (e.g., rejection, social exclusion, victimization, conflict, friendlessness) can shape developing biological stress-response systems and consequent adaptation. Indeed, several social-psychological models provide a framework for predicting how peer adversity may influence stress-response systems and associated physiological processes. These theories have in common the notion that humans have a fundamental need for relatedness that scaffolds their attention to, encoding and interpretation of, and reactions to social cues. Thus, exposure to peer adversity may impact the development of stress-response systems, calibrating them toward heightened or dampened sensitivity to social rewards (e.g., acceptance, positive feedback) and social punishments (e.g., rejection, negative feedback).

The need to belong theory (Baumeister & Leary, 1995) proposes that humans have a fundamental drive to develop and maintain stable, positive relationships. This need is believed to (a) guide cognitive processing in an effort to monitor one's current level of belonging and alert one to a mismatch between one's current and desired state; (b) exert positive affective consequences upon detection of fulfilled needs (e.g., presence of positive social bonds) and negative affective consequences upon detection of unmet needs (e.g., presence of negative social bonds or social deprivation); (c) motivate goal-oriented behavior aimed at achieving a sense of belonging; and (d) foster lasting adverse effects on health and adaptation when needs are unmet.

Building on this theory, Gardner and colleagues (2000) and Pickett et al. (2004) suggest that belonging needs are regulated by a social monitoring system that guides social information processing and detects the extent to which belonging needs are met. Similarly, Leary and colleagues (Leary et al., 1995) propose that self-esteem serves as a sociometer, indicating an individual's current level of social acceptance. Unmet belonging needs are presumed to arouse "social hunger" (Gardner et al., 2000), thereby heightening vigilance (e.g., increased attention and memory) to both positive and negative socially relevant cues aimed at determining opportunities for belonging, as well as "social snacking" behavior aimed at restoring one's belonging (Gardner et al., 2000).

Exposure to peer adversity, including both a lack of social reward (e.g., low acceptance, friendlessness) and the presence of social punishment (e.g., victimization, rejection), creates a disequilibrium between belonging needs and children's current social state, which likely alerts the social monitoring system and promotes sensitivity to both rewarding and aversive social cues. Either directly or via the social monitoring system, threats to belonging also may activate biological "alarm" systems, including neural circuitry involved in processing pain (Eisenberger, 2012) as well as stress-response systems (HPA and sympathetic-adrenal-medullary axis) and pro-inflammatory processes (Slavich et al., 2010). Consistent with developmental theories regarding the impact of early experience on stress physiology, disruption in these systems may be reflected in either upregulation (i.e., a vigilant pattern) or downregulation (i.e., an unemotional pattern), depending on features of the stressors, the individuals, and the developmental stage at which adversity occurs.

Summary Social-psychological theories suggest that the need to belong is a universal human drive that underlies multiple psychological processes (e.g., motivation, cognition, emotion) involved in social information processing. When this need is thwarted (e.g., through acute rejection), sensitivity to social cues that indicate belongingness (or lack thereof) increases. Although the need to belong is thought to be universal, there are individual differences in its intensity (Pickett et al., 2004), perhaps resulting in part from past social experiences. Exposure to peer adversity, which is characterized by negative social contacts and lack of stable, satisfying relationships, likely creates unsatiated or "socially hungry" (Gardner et al., 2000) children, who are particularly attuned, both cognitively and biologically, to cues involving social reward (e.g., group belonging; Telzer et al., 2019) and punishment (e.g., social exclusion; Rudolph et al., 2016b). Thus, peer adversity may serve a formative role in shaping individual differences in the social monitoring system and biological stress-response systems in ways that impact future sensitivity to social contexts.

Intersection of Theoretical Frameworks with Stress-Response Systems

Although each component of the stress-response systems is responsive to a range of stressors, they are especially sensitive to social-evaluative stressors that threaten social identity and belonging (Slavich et al., 2010). A meta-analysis reveals that social-evaluative threat, especially in the context of an uncontrollable outcome, predicts the largest and most reliable cortisol response as well as the slowest recovery to baseline (Dickerson & Kemeny, 2004). Similarly, the sympathetic-adrenal-medullary axis, which releases epinephrine and norepinephrine (catecholamines involved in the "fight-or-flight" response of the SNS) and guides CV response, shows sensitivity to social-evaluative threat (Mendes et al., 2008; Slavich et al., 2010). Recent research also suggests that the immune system is particularly sensitive to social adversity (Slavich & Cole, 2013; Slavich et al., 2010), perhaps because historically social conflict or rejection could heighten physical danger due to injury or isolation; (Slavich & Cole, 2013; Slavich & Irwin, 2014). Specifically, acute social stress predicts antiviral defenses (Edwards et al., 2006), whereas prolonged social stress predicts a reduction in such responses (Irwin et al., 2013). These responses are helpful in combatting injuries associated with physical threats (e.g., Irwin & Cole, 2011) but may increase risk for infection or inflammation-related disease (Slavich & Irwin, 2014) in the face of social threats.

Peer adversity often is characterized by social evaluation (e.g., teasing) and threats to belongingness (e.g., rejection, exclusion) and social identity (e.g., loss of close friendships), making it likely to trigger activation of the social monitoring system (Gardner et al., 2000). The HPA axis, SNS, and immune system may serve as biomarkers of the social monitoring system, activating when an imbalance between belongingness needs and one's current state is detected. Consistent with biological sensitivity to context theory and adaptive calibration model, peer adversity may calibrate stress-response systems, promoting either sensitization (e.g., a vigilant profile) or desensitization (e.g., an unemotional profile) that is reflected in hyper- or hypoactivation in the face of future stressors as well as upregulated or dampened basal function (Ellis et al., 2017).

Empirical Evidence for Peer Adversity Effects

Although many studies examine the impact of family adversity on the development of stress-response systems, less is known about the long-term impact of peer adversity. However, an emerging body of research explores how recent peer experiences influence the short- and long-term functioning of stress-response systems by calibrating them to be hyper- or hyporesponsive to subsequent social threats.

Hypothalamic-Pituitary-Adrenal (HPA) Axis

Research examining the effects of peer adversity on HPA axis function considers indexes of basal function (CAR; diurnal slope) as well as reactivity and regulation in the context of lab-based stressors (post- minus pre-stressor difference; AUC). Typically, studies of stress reactivity involve exposing youth to social stressors in the lab, which tend to evoke adaptive increases in cortisol reactivity, although there are individual differences in the level of reactivity (Blackhart et al., 2007; Dickerson & Kemeny, 2004). In line with theoretical perspectives highlighting the context specificity of the influence of early adversity on developmental adaptation (e.g., Ellis et al., 2017; Juster et al., 2010; O'Connor, 2003), these studies vary widely in terms of measurement of peer adversity (e.g., timing, severity, duration), type of lab-based stressor, index of activation (e.g., anticipatory, reactivity, regulation), and characteristics of the samples (e.g., age, risk status). Table 1 presents a summary of studies focused on the HPA axis.

Basal/Diurnal (Tonic) Variation Studies examining the effects of peer adversity on basal HPA activity often suggest that diurnal cortisol levels are dysregulated for children who face peer adversity. Elementary school students who experience high peer exclusion or low peer acceptance (fewer peer nominations of liking) show elevated AUC cortisol during school (Peters et al., 2011) and at home (Behnsen et al., 2019). In contrast, in late elementary to middle schoolers, peer victimization is significantly associated with lower basal cortisol at home (Kliewer, 2006; Vaillancourt et al., 2008). However, the same studies by Behnsen et al. (2019) and Peters et al. (2011) found no significant associations between high peer nonacceptance (more peer nominations of disliking) or victimization and AUC cortisol. Finally, no significant associations were found between bullying and basal morning cortisol collected at school in a study of sixth graders (Carney et al., 2010). These

Table 1 Exposure to peer adversity and the hypothalamic-pituitary-adrenal (HPA) axis

Study	Sample characteristics	Peer adversity	Lab stressor	Basal activity	HPA reactivity/regulation	Pattern of activation
Peer victimization						
Brendgen et al. (2017b)[1]	$N = 136$ MZ twin pairs; 54% female; followed fourth to eighth grade	SEQ	None	Bullied twin → steeper decrease in cortisol throughout the day (4 assessment points: 2 morning, 1 afternoon, 1 evening)		Hypo
Calhoun et al. $(2014)_a$	$N = 62$; 73% female; $M_{age} = 15$	Revised PEQ	TSST		Relational victimization → lower cortisol reactivity and faster recovery	Hypo
Kliewer $(2016)^2{}_b$	$N = 242$; 57% female; $M_{age} = 12$; followed across 3 years	SCEV	Social competence interview		Victimization → lower AUC cortisol across the interview (wave 4 assessment)	Hypo
Knack et al. $(2011)_c$	$N = 107$; 56% female; $M_{age} = 12$	CSEQ (self- and parent report); DIAS (self- and parent report)	Modified TSST	Victimized group → lower CAR; flatter cortisol decline (4 assessment points: 2 morning and 2 evening)	Victimized group → immediate post-task to 30 minute post-task cortisol decrease Non-victimized group → immediate post- task to 30 minutes post-task cortisol increase	Hypo

Ouellet-Morin et al. (2011, 2013)	$N = 30$ MZ twin pairs; 57% female; $M_{age} = 12$; followed ages 7–12	Mother report at all time points and self-report at 12 years	Modified TSST at 12 years		Non-bullied twins → post-task cortisol increase Bullied twins → no post-task increase More bullying → lower AUC cortisol secretion across the task	Hypo
Vaillancourt et al. (2008)[3]d	$N = 154$; 51% female; $M_{age} = 147$ months	Peer victimization questionnaire	None	Verbal victimization → lower basal cortisol across the day (2 assessment points: 1 morning and 1 evening)		Hypo
Chen et al. (2018)	$N = 80$; 45% female; fourth to fifth grade	OBVQ	Modified TSST		Victimization → higher cortisol levels at all time points Victimization → blunted decline from 25 minutes post-task to 40 minutes post-task relative to sharper decrease in controls	Hyper
Kliewer (2006)e	$N = 101$; 45% female; $M_{age} = 11$; high-risk neighborhood	ARCS	Video clip depicting violence, followed by a discussion	Victimization → lower basal cortisol at home (3 assessment points: morning)	Victimization → pre-task to post-task cortisol increase	Mixed

(continued)

Table 1 (continued)

Study	Sample characteristics	Peer adversity	Lab stressor	Basal activity	HPA reactivity/regulation	Pattern of activation
Brendgen et al. (2017a)[1]	159 MZ twin pairs and 120 DZ twin pairs; 52% female; $M_{age} = 14$	SEQ	None	No victimization effects on morning cortisol (CAR)		Null
Carney et al. $(2010)_f$	$N = 91$; 60% female; $M_{age} = 12$	EBE; SBS	None	Bullying not associated with basal cortisol (2 assessment points: beginning of school day and prior to lunch) or cortisol secretion across the morning		Null
Behnsen et al. $(2019)_g$	$N = 222$; 45% female, $M_{age} = 6.97$	SEQ – relational victimization	None	No effects of relational victimization on CAR, AUC, or diurnal cortisol slope		Null
Kliewer et al. (2012)[2]	$N = 228$; 55% female; $M_{age} = 12$; followed across 2 years	SEQ	SCI	No victimization/aggression effects on pre-task cortisol (wave 3 assessment)	No victimization/aggression effects on cortisol reactivity (wave 3 assessment)	Null
Knack et al. $(2011)_c$	$N = 107$; 56% female; $M_{age} = 12$	CSEQ (self- and parent report); DIAS (self- and parent report)	Modified TSST		No victimization effects on pre-task to immediate post-task cortisol change	Null
Peters et al. $(2011)_h$	$N = 97$; 47% female; $M_{age} = 9$	Sociometric nominations; FQQ	None	No victimization effects on AUC or diurnal cortisol slope (5 assessment points at home and school: 2 morning, 2 afternoon, and 1 evening)		Null

Rudolph et al. (2010, 2011)	$N = 132$; 52% female; $M_{age} = 9$	SEQ-R	Social challenge task	No victimization effects on pre-task cortisol	No victimization effects on 20 minutes post-task cortisol or pre-task to post-task cortisol change	Null
Vaillancourt et al. (2011)[3]	$N = 134$; 46% female; $M_{age} = 147$ months; followed for over 1 year	Peer victimization questionnaire	None	No victimization effects on basal cortisol at home (2 assessment points: 1 morning and 1 evening)		Null
Von Klitzing et al. (2012)	$N = 166$; 43% female; ages 5–6	PVS (teacher report); PVS of Berkeley puppet interview (child report)	Structured story completion task	No victimization effects on pre-task cortisol	No victimization effects on pre-task to 30 minutes post-task cortisol change	Null
Williams et al. (2017)[4]	$N = 143$; 66% female; $M_{age} = 14$	PECK	None	No bullying effects on diurnal cortisol (2 assessment points: 1 morning and 1 afternoon)		Null
Williams et al. (2017)[4]	$N = 31$; 48% female; $M_{age} = 14$	PECK	None	No bullying effects on diurnal cortisol (2 assessment points: 1 morning and 1 afternoon)		Null
Friendship						
Calhoun et al. (2014)[a]	$N = 62$; 73% female; $M_{age} = 15$	Revised PEQ	TSST		Negative friendship quality → slower cortisol recovery	Hyper

(continued)

Table 1 (continued)

Study	Sample characteristics	Peer adversity	Lab stressor	Basal activity	HPA reactivity/regulation	Pattern of activation
Peer status						
Behnsen et al. $(2019)_g$	$N = 222$; 45% female, $M_{age} = 6.97$	Peer nominations of liking (peer acceptance)	None	Low peer acceptance → higher CAR; higher AUC, flatter cortisol decline across the day		Hyper
Peters et al. $(2011)_h$	$N = 97$; 47% female; $M_{age} = 9$	Sociometric nominations; FQQ	None	Exclusion → higher AUC at school; lower wake-up cortisol; flatter cortisol decline across the day		Mixed
Behnsen et al. $(2019)_g$	$N = 222$; 45% female, $M_{age} = 6.97$	Peer nominations of disliking (nonacceptance)	None	No effects of peer nonacceptance on CAR, AUC, or diurnal cortisol slope		Null
Community violence						
Murali and Chen $(2005)_i$	$N = 115$; 62% female; ages 16–19	Exposure to violence interview	Debate or verbal puzzle with experimenter	Frequency of violence experience → higher pre-task cortisol		Hyper
Murali and Chen $(2005)_i$	$N = 115$; 62% female; ages 16–19	Exposure to violence interview	Debate or verbal puzzle with experimenter	No effects of total violence exposure on pre-task cortisol	No effects of total violence exposure on pre-task to 20 minutes post-task cortisol change	Null

Peer exclusion × friendship						
Peters et al. (2011)$_h$	$N = 97$; 47% female; $M_{age} = 9$	Sociometric nominations; FQQ	None	Exclusion × more friends → higher school AUC Exclusion × fewer friends → lower wake-up cortisol, flatter diurnal slope Exclusion × lower friendship quality → lower wake-up cortisol, flatter diurnal slope		Mixed
Friendship × responsiveness						
Calhoun et al. (2014)$_a$	$N = 62$; 73% female; $M_{age} = 15$	Revised PEQ	TSST		Low friendship quality × low responsiveness → slower post-task cortisol recovery High friendship quality × high responsiveness → slower post-task cortisol recovery	Hyper
Individual differences						
Carney et al. (2010)$_f$	$N = 91$; 60.4% female; $M_{age} = 12$	EBE; SBS	None	Bullying exposure × high anxiety → lower cortisol secretion across the morning at school		Hypo
Kliewer (2016)$_b$	$N = 242$; 57% female; $M_{age} = 12$	SCEV	Social competence interview		Victimization × poor emotion regulation → lower AUC cortisol across the interview	Hypo

(continued)

Table 1 (continued)

Study	Sample characteristics	Peer adversity	Lab stressor	Basal activity	HPA reactivity/regulation	Pattern of activation
Gender differences						
Vaillancourt et al. (2008)$_d$	$N = 154$; 51% female; $M_{age} = 147$ months	Peer victimization questionnaire	None	Verbal victimization → higher basal cortisol across day for boys and lower basal cortisol across day for girls		Mixed (girls: Hypo; boys: hyper)
Kliewer (2006)$_e$	$N = 101$; 45% female; $M_{age} = 11$; high-risk neighborhood	SCEV	Video clip depicting violence, followed by a discussion	Exposure to violence → atypical CAR pattern in girls, not boys Exposure to violence → lower pre-task cortisol in boys than girls	Exposure to violence → pre-task to 20 minutes post-task cortisol increase in boys, not girls	Mixed (boys)

Note. Subscripts indicate results from the same article. Superscripts indicate results published from the same/a subset of the same study sample. Brendgen et al. (2017a), Kliewer (2016), Kliewer et al. (2012), Ouellet-Morin et al. (2011), and Ouellet-Morin et al. (2013) assessed peer victimization longitudinally. *ARCS* adolescent resource challenges scale, *DIAS* direct and indirect aggression scales-victim version, *EBE* exposure to bullying events, *FQQ* friendship quality questionnaire, *OBVQ* Olweus bully/victim questionnaire, *PECK* personal experiences checklist, *PEQ* peer experiences questionnaire, *PVS* peer victimization scale, *SACAA* stigma against children affected by AIDS, *SBS* school bullying survey, *SCEV* survey of children's exposure to violence. Measures of peer adversity were self-report unless otherwise noted. [#]Additional null effects of gender are reported in the text divergent findings may be due to differences in developmental stages or the type of peer adversity.

Examining patterns of cortisol change across the day, Behnsen and colleagues (Behnsen et al., 2019) found that low peer acceptance predicted heightened CAR (difference between 30 minutes post-wake-up minus awakening level) and smaller declines throughout the day, suggesting a flattened diurnal response. Two other studies revealed that excluded 9-year-olds (Peters et al., 2011) and victimized adolescents (Knack et al., 2011) had lower than normal levels of morning cortisol (measured directly after waking; Peters et al., 2011) and CAR (difference between 30 minutes post-wake-up minus awakening level; Knack et al., 2011) and a flattened diurnal response across the day. In contrast, a study examining eighth-grade monozygotic and dizygotic twin pairs revealed that twins who were more victimized than their co-twin showed a steeper decline in cortisol across the day, controlling for CAR (Brendgen et al., 2017b). Yet other studies reveal nonsignificant correlations between peer adversity and basal/diurnal cortisol levels. For instance, two studies of younger children found nonsignificant effects of peer nonacceptance and victimization (Behnsen et al., 2019; Peters et al., 2011) on diurnal cortisol slopes. Another longitudinal study of 12-year-olds revealed a nonsignificant main effect of peer victimization on morning and evening cortisol levels concurrently and after an 18-month period (Vaillancourt et al., 2011). Other studies found that peer victimization in 14-year-olds was not significantly correlated with CAR (Brendgen et al., 2017a) or diurnal cortisol (Williams, 2017; Williams et al., 2017). However, in a study of late adolescents, frequency of exposure to community violence was significantly correlated with higher baseline cortisol levels in the lab, although total exposure (frequency, proximity, and severity) to community violence was not (Murali & Chen, 2005).

Cortisol Reactivity Several studies of victimized youth reveal attenuated cortisol secretion in response to acute in vivo social stressors. In one study examining the prospective effects of victimization in urban fifth and eighth graders, severe victimization (e.g., being physically threatened or harmed) over the past 3 years predicted lower total cortisol output (AUC) during a stress task (recalling and discussing a stressful event) 1 year later (Kliewer, 2016).[1] Similarly, another study of adolescents found that relational victimization over the past year was associated with less cortisol secretion to a social stressor (Calhoun et al., 2014). Exposure to victimization in fifth to eighth graders also was correlated with a decrease in cortisol from post social stressor to approximately 30 minutes following the stressor, compared to a continued increase in non-victimized youth (Knack et al., 2011). Prospective effects of victimization (measured at ages 7, 10, and 12) on lower cortisol reactivity also emerged in a sample of 12-year-old monozygotic twins who were discordant on bullying-victimization experiences. Whereas non-bullied twins exhibited increases

[1] Using physiological measures from a different time point, an overlapping sample showed nonsignificant correlations between physical/relational victimization and cortisol levels before/during/after the task (Kliewer et al., 2012)

in cortisol levels 25–30 minutes after a social stressor, their bullied co-twins did not (Ouellet-Morin et al., 2011; Ouellet-Morin et al., 2013). Furthermore, within the group of victimized co-twins, those with more frequent, severe, and chronic bullying experiences had lower AUC cortisol secretion during the lab stressor (Ouellet-Morin et al., 2011).

In contrast, a few studies link exposure to peer adversity with heightened cortisol reactivity in response to stress. One study in a younger (10-year-old) sample of Chinese children found that although all children showed a rapid increase and progressive decline in cortisol levels during a social-evaluative stressor, bullied children had higher cortisol levels at all measured time points compared to non-bullied children (Chen et al., 2018). Another study of African-American violence-exposed youth (age 11) found that recent peer victimization was associated with increases in cortisol immediately after viewing a video depicting community violence (Kliewer, 2006). However, a study of late adolescents revealed that experiencing community violence was not significantly correlated with cortisol reactivity to an acute social stressor (Murali & Chen, 2005). Other studies also found nonsignificant effects of peer adversity on cortisol reactivity. For example, peer victimization in third graders (Rudolph et al., 2010, 2011) and fifth to eighth graders (Knack et al., 2011) did not predict cortisol reactivity across social stressors. Another study also reported nonsignificant effects of peer victimization on concurrent and prospective (1 year later) cortisol levels (baseline and 30 minutes post-task) within a sample of 5-year-olds (von Klitzing et al., 2012).

Cortisol Regulation Although less prevalent, a few studies examined the effect of peer adversity on post-stressor cortisol recovery. In adolescents, past-year relational victimization predicted less cortisol secretion (greater recovery) 40 minutes after a social stressor (although this effect was due to initial blunted reactivity), and poor friendship quality predicted reduced cortisol recovery (slower return to baseline; Calhoun et al., 2014). A study of 10-year-olds revealed that recently bullied youth showed more cortisol secretion 40 minutes following a social stressor relative to non-bullied youth (Chen et al., 2018).

Individual Differences Some research suggests the effect of peer adversity on HPA function differs contingent on individual differences in youth and their contexts. In adolescents, elevated victimization over 3 years predicted lower AUC cortisol in youth with poor relative to strong emotion regulation (Kliewer, 2016). In a sample of 9-year-olds, having fewer friends and lower friendship quality worsened the effect of peer exclusion on wake-up cortisol and diurnal response (resulting in lower initial levels and a flatter slope across the day rather than the typical decline), whereas higher friendship quality and having more friends countered this effect (Peters et al., 2011).[2] Furthermore, the association between exclusion and elevated

[2]These findings reflect a moderation of the main effect reported in the basal/diurnal variation section.

AUC cortisol across the school day was stronger for children with more than fewer friends (Peters et al., 2011).[3] Another study of adolescents (ages 12–16) examined the moderating effects of post-stressor friendship support (responsiveness) on cortisol recovery after a social stressor (Calhoun et al., 2014). A two-way interaction between positive friendship quality and post-stressor friendship support revealed that adolescents with poorer-quality friendships showed slower cortisol recovery when their friends offered less than more support, whereas adolescents with higher-quality friendships showed slower cortisol recovery when their friends offered more than less support (Calhoun et al., 2014). Observed adverse effects of friendship number and quality on HPA reactivity and regulation may reflect effects of co-rumination with friends (Calhoun et al., 2014; Peters et al., 2011), an issue that warrants further investigation Finally, greater exposure to bullying predicted lower morning and afternoon cortisol levels in sixth graders with high relative to low levels of anxiety (Carney et al., 2010).[4]

Gender Differences Studies also have examined whether gender moderates the effect of peer adversity on HPA activity. In 12-year-olds, Vaillancourt et al. (2008)[5] reported that verbal victimization predicted higher diurnal cortisol secretion in boys but lower diurnal cortisol secretion in girls. Kliewer (2006) found that exposure to violence predicted lower pre-task cortisol and higher post-task cortisol reactivity in boys but not in girls. In contrast, more exposure to violence predicted atypical CAR (cortisol decrease instead of increase from wake-up to 30 minutes later) in girls but not in boys (Kliewer, 2006). Yet, many studies fail to document gender moderation effects (Brendgen et al., 2017b; Kliewer, 2016; Ouellet-Morin et al., 2011; Rudolph et al., 2010, 2011; von Klitzing et al., 2012), suggesting further research is needed.

Summary Although research supports an association between peer adversity and HPA dysregulation, different patterns emerge across studies. Studies examining basal/diurnal variation indicate evidence for nonsignificant, attenuated, and heightened diurnal cortisol in youth with a history of peer adversity, whereas findings for studies examining cortisol reactivity are less balanced. Consistent with an "unemotional" stress-response profile (Ellis et al., 2017), many studies reveal lower cortisol reactivity to stress in youth with than without a recent history of peer adversity. However, a few studies reveal a "vigilant" stress-response system profile wherein peer adversity is linked to heightened cortisol reactivity to stress. Moreover, three studies failed to find any significant main effects, although for one study, HPA activity moderated the effects of peer adversity on later symptoms of psychopathology (Rudolph et al., 2010, 2011), providing evidence for cortisol reactivity as a form of

[3] These findings reflect a moderation of the main effect reported in the basal/diurnal variation section.

[4] These findings reflect a moderation of the main effect reported in the basal/diurnal variation section.

[5] These findings reflect a moderation of the main effect reported in the basal/diurnal variation section.

biological sensitivity to context. Variability in findings may be attributable to differences in methods, such as the setting (home, school, lab) or the timing and frequency of samples, or may reflect age differences (Miller et al., 2007). Changes in cortisol activity may occur with development given physiological and social changes during adolescence or may be a function of adolescents having more years of exposure to adversity. Indeed, an attenuated cortisol response following longer exposure (e.g., Kliewer, 2016) may reflect a long-term adaptation to chronic stress and initial greater secretion of cortisol (Juster et al., 2010) or exposure to more severe stressors that result in desensitization of stress-response systems (Ellis et al., 2017).

Autonomic Nervous System (ANS)

Research has explored how peer adversity leads to individual differences in resting ANS function as well as SNS reactivity and PNS regulation following lab-based stressors. These stressors typically evoke SNS reactivity, as reflected in increases in heart rate, blood pressure, skin conductance, and levels of sAA, which then prompt an increase in PNS regulation, reflected in elevated HRV and RSA, in order to bring ANS function back to baseline. As with research on the HPA axis, studies vary considerably in terms of approaches to measuring peer adversity (e.g., victimization, exposure to violence, low peer approval) and ANS function (e.g., cardiovascular activity, skin conductance, digestive enzymes) and reveal different effects of adversity over time and across developmental stages. Table 2 presents a summary of studies focused on the ANS.

Resting ANS Function Studies examining ANS activity at rest typically do not separate peer victimization from other forms of adversity (e.g., violence exposure), and research on peer adversity and resting ANS function in elementary school children is limited. Overall, available research reveals heightened activation of the SNS in individuals with a history of peer adversity, but more mixed findings regarding PNS regulation. With regard to SNS activation, African-American adolescent males with heightened exposure to peer victimization (e.g., being beaten up) as well as community violence (e.g., experiencing or witnessing a mugging) within the past year showed a smaller day to night shift in blood pressure (Wilson et al., 2002). This effect was driven by elevated nighttime blood pressure (which is thought to reflect SNS hyperactivation) in adversity-exposed youth. In another study of high schoolers, more lifetime exposure to violence (school, home, and neighborhood), regardless of whether it was perpetrated by peers, was associated with elevated systolic and diastolic blood pressure (SBP, DBP) and heart rate (HR) at rest, although no effect of violence exposure was found for basal HRV, a marker of PNS activity (Murali & Chen, 2005). One study of 5–10-year-olds found that current peer problems were associated with less effective parasympathetic regulation of heart rate variability (HRV) at rest (Michels et al., 2013). In contrast, a study of emerging adults found that retrospectively reported high school peer victimization was not associated with

Table 2 Exposure to peer adversity and the autonomic nervous system (ANS)

Study	Sample characteristics	Peer adversity	Lab stressor	Resting ANS	ANS reactivity (SNS)	ANS regulation (PNS)	Pattern of activation
Peer victimization							
Iffland et al. (2014)$_j$	$N = 74$; 59% female; $M_{age} = 24$; one-half had social anxiety diagnosis	SSEQ	Cyberball		Victimization → attenuated SCL increase from inclusion to exclusion		Hypo SNS
Iffland et al. (2018)	$N = 94$; 64.9% female; $M_{age} = 25$	SSEQ	Negative or neutral evaluative videos		Victimization → higher HR during videos		Hyper SNS
Kliewer et al. (2012)$_e$	$N = 228$; 55% female; $M_{age} = 14$	SEQ	SCI		Victimization → higher sAA at the end of interview		Hyper SNS
Mazzone et al. (2017)	$N = 28$; 39% female; $M_{age} = 12$	Peer nominations	Cyberball		Victimization → higher skin nose temperature to exclusion (but not inclusion)		Hyper SNS
Bollmer et al. (2006)	$N = 99$; 49% female; $M_{age} = 11$	PPS (parent and self-report)	Victimization narrative		No victimization effects on SCL during the victimization narrative relative to baseline		Null
Breslend et al. (2018)	$N = 119$; 100% female; $M_{age} = 12$	SEQ (counselor report)	Modified SCI		No victimization effects on SCL during interview relative to baseline	No victimization effects on RSA during interview relative to baseline	Null
Erath et al. (2018)[1]; Gregson et al. (2014)[1]	$N = 123$; 50% female; $M_{age} = 12$	SEQ (self-report); SBSR (teacher report)	TSST		No victimization effects on SCL during TSST controlling for baseline		Null

(continued)

Table 2 (continued)

Study	Sample characteristics	Peer adversity	Lab stressor	Resting ANS	ANS reactivity (SNS)	ANS regulation (PNS)	Pattern of activation
Iffland et al. (2014)$_j$	$N = 74$; 59% female; $M_{age} = 24$; one-half had social anxiety diagnosis	SSEQ	Cyberball		No victimization effects on HR		Null
Murray-Close et al. (2014)	$N = 196$; 54% female; $M_{age} = 10$	CSES (teacher report)	SCI		No victimization effects on SBP, DBP, or SCL during interview relative to baseline		Null
Newman (2014)	$N = 57$; 70% female; $M_{age} = 21$	EBQ	Exclusion from online chat		No victimization effects on PEP during exclusion	No victimization effects on RSA during exclusion	Null
Rudolph et al. (2010, 2011)[2]	$N = 132$; 52% female; $M_{age} = 9$	SEQ-R	Social challenge task	No victimization effects on pre-task sAA	No victimization effects on 20 minutes post-task sAA or pre-task to post-task sAA change		Null
General peer adversity							
Michels et al. (2013)	$N = 334$; 49% female; ages 5–10	SDQ (parent report)	None	Peer adversities → less effective HRV regulation while resting			Hypo PNS
Erath and Tu (2014)$_b$	$N = 123$; 50% female; $M_{age} = 12$	TCPR	TSST			Lower social competence → less RSA regulation during task relative to baseline	Hypo PNS
Erath and Tu (2014)$_b$	$N = 123$; 50% female; $M_{age} = 12$	TCPR	TSST		No social competence effects on SCL		Null

Friendship/peer status							
Ponzi et al. (2016)	$N = 44$; 48% female; $M_{age} = 8$	Friendship network density	Interview about peer network		Lower friendship network density → elevated pre-task to post-task sAA increase		Hyper SNS
Graziano et al. (2007)$_g$	$N = 341$; 54% female; $M_{age} = 6$	Peer nominations	LAB-TAB cognitive tasks			Lower peer status → less RSA suppression during tasks controlling for baseline	Hypo PNS
Community violence							
Wilson et al. (2002)$_d$	$N = 56$; 52% female; $M_{age} = 13$	SECV	None	Violence → smaller day-night shift in BP			Hyper SNS
Murali and Chen (2005)$_a$	$N = 115$; 62% female; $M_{age} = 17$	Exposure to violence interview	Debate or verbal puzzle with experimenter	Violence → higher resting SBP, DBP, HR	Violence → lower SBP, DBP, HR during interview controlling for baseline	Violence → attenuated HRV during interview controlling for baseline	Mixed
Murali and Chen (2005)$_a$	$N = 115$; 62% female; $M_{age} = 17$	Exposure to violence interview	Debate or verbal puzzle with experimenter	No violence effects on resting HRV			Null
Individual differences							
Erath et al. (2012)[1]$_f$	$N = 63$; 49% female; fifth to sixth grade	SEQ (self- and parent report)	TSST		Victimization → higher HR during task for socially anxious youth no victimization effects on SCL	Victimization → less RSA regulation during task in socially anxious youth	Hyper SNS Hypo PNS

(continued)

Table 2 (continued)

Study	Sample characteristics	Peer adversity	Lab stressor	Resting ANS	ANS reactivity (SNS)	ANS regulation (PNS)	Pattern of activation
Monti et al. (2014)[2]	$N = 118$; 52% female; $M_{age} = 9$	SEQ	Social challenge task		Victimization → pre-task to post-task sAA increase in youth whose mothers encouraged less engagement coping		Hyper SNS
Gender differences							
Hamilton et al. (2008)$_c$	$N = 93$; 57% female; $M_{age} = 19$; one-half elevated past victimization	EBQ	TSST speech preparation		Past victimization → blunted BP and HR during task in males, not females		Hypo SNS
Murali and Chen (2005)$_a$	$N = 115$; 62% female; $M_{age} = 17$	Exposure to violence interview	Debate or verbal puzzle with experimenter			Violence → higher HRV during task, controlling for baseline, in boys, not girls	Hypo PNS
Aults et al. (2019)	$N = 82$; 52% female; $M_{age} = 12$	Peer nominations	Aversive sound	Victimization → higher resting HR in boys, not girls			Hyper SNS

Note. Subscripts indicate results from the same article. Superscripts indicate results published from the same/a subset of the same studied sample. See Gregson et al. (2014) for null sex difference finding. *CSES* children's social experience scale, *EBQ* experiences with bullying questionnaire, *PPS* perception of peer support scale, *SBRS* social behavior rating scale, *SCI* social competence interview, *SDQ* strengths and difficulties questionnaire, *SECV* survey of exposure to community violence, *SEQ* social experiences questionnaire, *SSEQ* stressful social experiences questionnaire, *TCPR* teacher checklist of peer relations. Measures of peer adversity were self-report unless otherwise noted pre-ejection period (PEP, a marker of SNS activity) or respiratory sinus arrhythmia (RSA, an index of PNS function) at rest (Newman, 2014). Thus, there is some evidence for disrupted resting PNS regulation, but results across studies are inconsistent, with the limited available support coming from younger samples.

ANS Reactivity Several studies examining variability in ANS reactivity to lab stressors suggest that peer adversity is associated with SNS hyperreactivity. For example, elementary school children who reported having fewer friends showed higher levels of sAA than those with more friends while discussing their peer networks (Ponzi et al., 2016). Adolescents with a history of peer victimization in the past month showed higher levels of sAA while talking about peer stressors (Kliewer et al., 2012). Recent victimization also was associated with higher nose skin temperature (reflecting SNS hyperreactivity) during a social exclusion task (Mazzone et al., 2017), suggesting that SNS hyperreactivity extends to contexts beyond those of the initial victimization. In adults, a history of peer victimization during childhood and adolescence predicted a higher heart rate when viewing videos of either neutral or negative evaluative statements (Iffland et al., 2018).

However, several studies reveal ANS hyporeactivity to an acute stressor in adolescents and adults with a history of peer adversity. In a retrospective study of adults who reported on exposure to peer victimization during childhood and adolescence, those with more past victimization showed an attenuated skin conductance response (thought to reflect dampened SNS reactivity) but no difference in HR reactivity during a social exclusion task (Iffland et al., 2014). In another retrospective study, adults with a history of peer victimization during high school showed a smaller increase in PEP during a task in which they were excluded from an online chat (Newman, 2014). In a study of adolescents exposed to peer and community violence, those with greater lifetime exposure to violence (e.g., experiencing or witnessing physical assaults) had higher baseline, as well as smaller increases in SBP, DBP, and HR during a stressful interaction task (Murali & Chen, 2005), indicating a dampened response to an acute stressor following exposure to adversity.

Despite research suggesting a link between exposure to peer adversity and atypical SNS reactivity to acute stressors, several studies reveal null effects. For example, some studies of early adolescents reveal no evidence for elevated skin conductance (Bollmer et al., 2006; Breslend et al., 2018) or blood pressure (Murray-Close et al., 2014) when peer-victimized youth report on previous conflicts with peers. Similarly, research failed to support elevated SCL when recently victimized early adolescents gave a speech that they were told would be evaluated by peers (Erath et al., 2018; Gregson et al., 2014), and Rudolph et al. (2010, 2011) did not find a main effect of prior peer victimization on elevated sAA in 9-year-olds engaging in a stressful peer interaction task.

ANS Regulation in Response to Stress Research links peer adversity and less supportive peer networks with poorer regulatory function of the PNS in response to stressors. Supporting the association between involvement with peers and PNS regulation, lower social competence (which included being less well-liked by classmates) was associated with less RSA suppression during a task in which fifth and sixth graders were told that their conversations with an experimenter were being evaluated by same-age peers (Erath & Tu, 2014). Similarly, peer-nominated high social status was associated with greater RSA suppression when young children completed a cognitive control task, suggesting that lower adversity is associated with more effective PNS regulation (Graziano et al., 2007). Extending findings to other forms of adversity, exposure to peer and community violence was linked to a smaller increase in HRV when adolescents engaged in a stressful interaction with an experimenter (Murali & Chen, 2005). However, in early adolescent girls, there was no link between concurrent victimization and RSA suppression when youth reported on negative past peer experiences (Breslend et al., 2018), suggesting variability in the link between peer adversity and PNS regulation.

Individual Differences A few studies reveal individual differences in the link between peer adversity and ANS function. During an evaluative social interaction task, more recent victimization in fifth and sixth graders was associated with elevated heart rate (but not skin conductance), but only when youth also had elevated levels of social anxiety symptoms (Erath et al., 2012). A study of third graders found that previous-year peer victimization was associated with higher sAA during a stressful social interaction in youth whose mothers were less encouraging of engagement coping (Monti et al., 2014). Suggesting important gender differences, victimization in mid-adolescents was associated with higher resting HR when boys but not girls tried to ignore an aversive sound (Aults et al., 2019). Similarly, a history of peer and community violence exposure predicted more HRV in adolescent boys but not girls (Murali & Chen, 2005). In contrast, adult men but not women with prior peer victimization showed an attenuated change in HR and blood pressure during a speech preparation task (Hamilton et al., 2008), suggesting that gender effects (i.e., more hypo- or hyperreactivity) may not be consistent across samples.

Summary To date, little research examines resting ANS function in youth exposed to peer adversity, particularly in elementary school. However, in line with allostatic load models, some studies reveal elevated resting SNS activity in victimized youth and in adolescents exposed to peer and community violence. Moreover, peer adversity was linked with poorer PNS activity at rest in one study of younger children, but limited research in adolescents and adults yields null effects. Consistent with stress-sensitization models, most studies examining ANS responses to lab stressors reveal that youth exposed to more peer adversity show elevated SNS reactivity and poorer PNS regulation, but some studies reveal hyporeactivity of the SNS in adults and adolescents, and several studies yielded null effects. The small amount of research exploring individual differences in the link between peer adversity and ANS function supports interactive models, finding that peer adversity predicts elevated SNS reactivity only in youth with concurrent psychological or environmental vulnerability.

One explanation for inconsistencies in the direction of effects (hyper- or hypoactivation following adversity) may be the sample age, as hyperactivation is more common in young children and adolescents and hypoactivation is more common in adults and older adolescents (c.f. Iffland et al., 2014). Another explanation for these differences is the chronicity of exposure. Studies examining recent exposure in youth typically find evidence of SNS hyperreactivity to stressors, whereas studies examining exposure to chronic victimization or community violence in adolescents and emerging adults typically find evidence of blunted SNS reactivity. These results conform to allostatic load models, which argue that adversity may lead to overarousal in the short term but to underarousal over time as the system's ability to respond is overtaxed by chronic strain (Juster et al., 2010). It is also possible that conflicting results arise from the fact that various aspects of ANS function respond differently to peer adversity.

Immune System

Studies investigating the contribution of peer adversity to immune system function have examined both basal systemic inflammation as well as inflammatory responses to acute stress. Functioning of the immune system has been captured in indexes of both pro-inflammatory signaling processes (i.e., changes in NF-kB and I-kB) and the products of these processes (e.g., pro-inflammatory cytokines, such as IL-6 and IL-1β) and inflammatory biomarkers reflecting systemic inflammation (e.g., C-reactive protein; CRP). Table 3 presents a summary of studies focused on the immune system.

Systemic Inflammation and Pro-inflammatory Signaling Research reveals that exposure to bullying in childhood and adolescence predicts greater systemic inflammation in the short and long term. Copeland et al. (2014) found that cumulative exposure to bullying during childhood/adolescence (ages 9–16) predicted elevated CRP during the same time period as well as into adulthood (ages 19–21). Similarly, Takizawa et al. (2015) found that frequent bullying at age 7 or 11 (or some bullying at both ages) predicted higher levels of CRP at age 45. Lower peer preference at age 13 also predicts higher levels of sensitive CRP (hsCRP) at age 16 (de Bruine et al., 2019).

Examining markers of pro-inflammatory signaling (NF-kB and I-kB) and biomarkers of systemic inflammation (CRP and IL-6), one study followed a sample of 15–19-year-old girls at risk for major depressive disorder, measuring targeted rejection and inflammation every 6 months over 2.5 years (Murphy et al., 2013). Compared to visits where recent rejection had not occurred, experiences of targeted rejection were associated with higher pro-inflammatory signaling but not systemic

Table 3 Exposure to peer adversity and the immune system

Study	Sample characteristics	Peer adversity	Lab stressor	Basal activity	Immune system reactivity/regulation	Pattern of activation
Peer victimization						
Copeland et al. (2014)	$N = 1309$; 53% female; ages 9–16; $N = 759$; 55% female; ages 19–21	CAPA (self- and parent report)	None	Cumulative bullying at ages 9–16 → higher CRP (9–16 and 19–21)		Hyper
Giletta et al. $(2018)_a$	$N = 157$; 100% female; $M_{age} = 15$	Revised PEQ (peer report)	Modified TSST	Victimization → higher pre-task TNF-α and latent pro-inflammatory phenotype	Victimization → pre-task to post-task IL-1β and IL-6 increases	Hyper
Takizawa et al. (2015)	$N = 7102$; 49% female; followed ages 7–45	Parent interview	None	Bullying in childhood → higher CRP and fibrinogen later in life		Hyper
Giletta et al. $(2018)_a$	$N = 157$; 100% female; $M_{age} = 15$	Revised PEQ (peer report)	Modified TSST		No victimization effects on TNF-α or latent pro-inflammatory phenotype changes	Null
Peer rejection/status						
de Bruine et al. $(2019)_c$	$N = 587$; 54.6% female; $M_{age} = 11$ (at baseline)	Peer nominations	None	Lower peer preference at age 13 → higher hsCRP at age 16		Hyper
Murphy et al. $(2013)_b$	$N = 147$; 100% female; $M_{age} = 17$	LSI	None	Targeted rejection → higher NF-kB and I-kB mRNA		Hyper
Murphy et al. $(2013)_b$	$N = 147$; 100% female; $M_{age} = 17$	LSI	None	No targeted rejection effects on CRP or IL-6		Null

Peer rejection × social status						
Murphy et al. (2013)$_b$	$N = 147$; 100% female; $M_{age} = 17$	LSI	None	Targeted rejection → higher NF-kB and I-kB mRNA as well as smaller ratio of NF-kB to I-kB mRNA in higher compared to lower social status girls		Hyper
Individual differences						
de Bruine et al. (2019)$_c$	$N = 587$; 54.6% female; $M_{age} = 11$ (at baseline)	Peer nominations	None	High peer popularity at age 13 × low early family adversity → highest hsCRP at age 16		Hyper
Giletta et al. (2018)$_a$	$N = 157$;100% female; $M_{age} = 15$	Revised PEQ (peer report)	Modified TSST		Victimization × high hopelessness → pre-task to post-task IL-1β, IL-6, TNF-α, and latent pro-inflammatory phenotype increases Victimization × negative cognitive style → no effect on pre-task to post-task cytokine changes	Hyper

Note. Subscripts indicate results from the same article. Copeland et al. (2014), Takizawa et al. (2015), and Murphy et al. (2013) assessed peer adversity longitudinally. *CAPA* child and adolescent psychiatric assessment, *LSI* life stress interview, *PEQ* peer experiences questionnaire. Measures of peer adversity were self-report unless otherwise noted inflammation. Taken together, studies link both early and recent adversity with elevated inflammation.

Immune System Reactivity One study supports the effect of peer adversity on immune system responses to acute laboratory stress. Giletta et al. (2018) examined the association between peer victimization and inflammatory response to a social-evaluative stressor in girls between ages 12 and 16 who were at risk for psychopathology. In line with previous work, those who had experienced more peer victimization showed higher levels of inflammation at baseline (for TNF-α but not IL-6 and IL-1β) and higher inflammatory responses (for IL-6 and IL-1β but not TNF-α) to the stressor, revealing an effect of peer adversity on the immune system at rest and in the face of acute stressors.

Individual Differences Examining whether perceived status in the peer group hierarchy moderated the association between targeted rejection and immune function, Murphy et al. (2013) found that girls with higher status showed greater inflammatory response (NF-kB and I-kB mRNA but not IL-6 or CRP) after rejection compared to those with lower status. de Bruine et al. (2019) found that higher peer popularity at age 13 predicted higher hsCRP at age 16 in adolescents with low but not average or high levels of early family adversity. Finally, Giletta et al. (2018) found that feelings of hopelessness (but not negative cognitive styles) moderated the link between peer victimization and inflammatory responses to a lab stressor, such that victimization was associated with increases in pro-inflammatory cytokines in girls who had high but not low levels of hopelessness.

Summary Although only a few studies examine the response of the immune system to peer adversity, greater inflammatory responses (both systemic and reactive) are seen across studies. Specifically, exposure to bullying and rejection is associated with increases in pro-inflammatory signaling months later, as well as increases in systemic inflammation months and years later. In at-risk adolescent girls, immune system responses in the face of lab stressors appear to be impacted by a history of peer adversity. Finally, individual differences in factors such as perceived social status and levels of hopelessness moderate the association between peer adversity and immune response. However to date, studies only explore a subset of populations, types of victimization, and immune responses. Much has yet to be determined regarding the influence of individual characteristics, such as gender, age, or chronicity of adversity, on the link between peer adversity and immune system function. Additional work exploring the impact of peer adversity on potential changes in the function of the immune system (e.g., pro-inflammatory signaling), instead of its final inflammatory outcome, may be especially insightful.

Future Directions for Research on Biological Embedding of Peer Adversity

Despite significant progress in understanding how peer adversity can "get under the skin" by influencing key biological stress-response systems, this research is still in the early stages. Here we delineate several directions for future research to elucidate how peer adversity calibrates developing stress-response systems and the implications for development.

Distinguishing Effects of Specific Types of Peer Adversity on Specific Indexes of Stress-Response Systems

Most studies in this area assess only one type of peer adversity (for exceptions, see Calhoun et al., 2014; Peters et al., 2011). Given the many methodological features that vary across studies (e.g., sample, timing of measurement, type of stress-response systems index, chronicity of adversity), it is difficult to compare the effect of different types of adversity across different studies. Thus, studies assessing multiple types of adversity will help to elucidate the specificity of peer adversity-linked stress-response systems variability. Moreover, most studies use self-reports of peer adversity; integrating ecologically valid assessments (e.g., peer nominations, observations) may inform future work. Similarly, many studies use single indexes of stress-response system functioning (e.g., either at rest or in response to stressors) despite some evidence for different patterns of findings across indexes (e.g., Brendgen et al., 2017a, b; Erath & Tu, 2014; Iffland et al., 2014; Kliewer, 2006; Murali & Chen, 2005; Peters et al., 2011). Furthermore, most studies assess stress-response system responses to lab-based social stressors (e.g., giving a speech). This approach is consistent with theory and research implicating the importance of social evaluation and threats to one's need to belong on stress-response system functioning; however, to confirm the specificity of these effects, it would be beneficial to compare responses to social vs. nonsocial stressors both in the lab and in natural settings.

Tracking Effects of Peer Adversity Over Time

Although theoretical frameworks of early adversity focus on understanding how stress-response systems adapt to perturbations in the environment (Ellis et al., 2017), much of the research is in adolescents, with less research examining elementary school students. Moreover, most research uses concurrent or short-term designs, and almost no research uses multiple assessments of stress-response system functioning to examine changes over time. This constraint limits our understanding in several ways. First, it is difficult to draw strong conclusions about the direction of effects. Although research in this area assumes a causal model, it is possible that youth who show strong biological reactivity are at greater risk for future disturbances in their peer relationships or that there is an incidental association between these two that is better accounted for by a shared risk factor (Rudolph & Asher, 2000). Second, despite potential differences in stress-response system profiles that may emerge in the short vs. long term, research makes it difficult to distinguish such profiles given limited data about the chronicity of peer adversity. Thus, future research will need to use prospective designs that assess frequency and severity of adversity and biological stress-response systems across multiple waves to compare the effects of early vs. recent vs. chronic adversity. Clearly distinguishing the severity and timing of adversity may help distinguish studies that reveal up- vs. down-regulation of stress-response systems. Moreover, the few studies that include longitudinal designs use between-subject analyses; research needs to track trajectories of stress-response systems over time using within-subject analyses to determine whether these systems move from a profile of hyperactivation (e.g., vigilance; Ellis et al., 2017) to hypoactivation (e.g., unemotional; Ellis et al., 2017) across development in the face of chronic adversity and whether profiles change in subsequent supportive contexts or across developmental transitions (e.g., recalibration at puberty; Del Giudice et al., 2011). Analytic approaches also need to consider the possibility of nonlinear associations between adversity and stress responses, which have rarely, if ever, been considered in peer adversity theory and research.

Exploring Individual, Gender, and Age Differences in the Effects of Peer Adversity

It is likely that inconsistent findings stem in part from individual differences in the impact of peer adversity on stress-response system functioning. Future research would benefit from considering qualities of youth (e.g., age, gender, temperament, cognitive styles, social goals) and their contexts (e.g., family support, school environment) that may determine the extent to which peer adversity alters stress-response system functioning and, perhaps, the nature of these alterations (e.g., heightened vs. attenuated activation). Moreover, there may be individual differences in the implications of stress-response system profiles for health. For example, heightened reactivity may be a risk factor for vigilant phenotypes but a protective factor for sensitive phenotypes; likewise, dampened reactivity may be a protective factor for buffered phenotypes but a risk factor for unemotional phenotypes (Ellis et al., 2017).

Linking Biological Embedding of Peer Adversity to Developmental Outcomes

Although the ultimate goal of research on biological embedding of peer adversity is to better understand developmental outcomes following adversity, few studies directly examine these pathways. Research links disrupted stress-response system functioning to multiple forms of psychopathology (e.g., anxiety, depression, aggression) and health difficulties (e.g., somatic symptoms, CV disease), providing indirect support for the implications of biological embedding. However, direct tests of mediation models are needed to determine specific pathways from certain types of peer adversity to certain types of stress-response system dysfunction to certain types of health outcomes. This approach also will clarify the health implications of different patterns of stress-response system functioning (e.g., heightened vs. attenuated activity). Research also needs to test theoretical perspectives (e.g. Ellis et al., 2011, 2017) suggesting that stress-response system changes in the context of adversity reflect developmental adaptations that serve a protective role, at least in the short term, and to better elucidate whether these same adaptations undermine health in the long term. Alternatively, perhaps biological sensitivity can be leveraged as a resource for responding to later improvements in relationships or direct intervention efforts. That is, biological sensitivity stemming from early peer adversity may reflect susceptibility to both positive and negative aspects of the environment. Directly testing these ideas would be highly informative for better understanding the developmental implications of stress-response system profiles following exposure to peer adversity and determining under what conditions they may reflect disturbances that portend future health costs versus adaptive advantages that confer future health benefits.

Integrating Genetics into Peer Adversity Models and Research

Refining models regarding the biological embedding of peer adversity will require a careful consideration of genetic contributions (see Brendgen & Kretschmer, current volume) to both exposure to peer adversity and emerging stress-response systems. Genetics may play several roles in models of peer adversity. On the one hand, some youth have a genetic liability to peer adversity exposure (Ball et al., 2008; Brendgen et al., 2011), perhaps reflected in a high-risk behavioral endophenotype (e.g., temperamental negative emotionality, behavioral inhibition, negative cognitive styles); the same genetic liability may contribute to stress-response system profiles, resulting in an incidental association between peer adversity exposure and developmental outcomes. Although some genetically informed research suggests that exposure to peer adversity (bullying) contributes to atypical stress responses independent of genetic makeup (Brendgen et al., 2017b; Ouellet-Morin et al., 2011), more research needs to address this potential confound. On the other hand, genetic liability may serve as an individual difference that helps to determine which youth are vulnerable to which stress-response system profiles (e.g., hyper- or hyporesponsive) or are protected against stress-response system disruption following peer adversity. Indeed, both molecular genetic (Benjet et al., 2010; DiLalla et al., 2015) and behavior genetic (Brendgen et al., 2008) research reveals that genetic liability moderates the impact of peer adversity on emotional and behavioral outcomes—it is possible that stress-response system functioning serves as a biological expression of this genetic liability. Finally, exposure to peer adversity may alter gene expression in ways that shape stress-response system development (Slavich & Irwin, 2014). Developing and testing comprehensive models of biological embedding of peer experience will require considering these multiple roles of genetic liability.

Conclusion

Given the alarming statistics regarding rates of exposure to peer adversity among youth, along with the multitude of adverse physical and mental health effects that persist into adulthood, it is critical to understand how these experiences become embedded in stress-response systems in ways that shape future developmental adaptation. A growing body of research supports the idea that exposure to peer adversity calibrates stress-response systems toward both hyper- and hyposensitivity in terms of both their tonic and stress-reactive functions. Future research should be directed toward better understanding the emergence and changes in stress-response system functioning across extended periods of time and determining why particular youth may follow diverse pathways of adaptation following peer adversity.

References

Aults, C. D., Machluf, K., Sellers, P. D., & Jones, N. A. (2019). Adolescent girls' biological sensitivity to context: Heart rate reactivity moderates the relationship between peer victimization and internalizing problems. *Evolutionary Psychological Science, 5*, 178–185.

Ball, H. A., Arseneault, L., Taylor, A., Maughan, B., Caspi, A., & Moffitt, T. E. (2008). Genetic and environmental influences on victims, bullies and bully-victims in childhood. *Journal of Child Psychology and Psychiatry, 49*(1), 104–112.

Barton, G. M. (2008). A calculated response: Control of inflammation by the innate immune system. *Journal of Clinical Investigation, 118*(2), 413–420.

Baumeister, R. F., & Leary, M. R. (1995). The need to belong: Desire for interpersonal attachments as a fundamental human motivation. *Psychological Bulletin, 117*(3), 497–529.

Beauchaine, T. P. (2015). Respiratory sinus arrhythmia: A transdiagnostic biomarker of emotion. dysregulation and psychopathology. *Current Opinion in Psychology, 1*(3), 43–47.

Behnsen, P., Buil, J. M., Koot, S., Huizink, A., & van Lier, P. (2019). Heart rate (variability) and the association between relational peer victimization and internalizing symptoms in elementary school children. *Development and Psychopathology, 32*(2), 521–529.

Benjet, C., Thompson, R. J., & Gotlib, I. H. (2010). 5-HTTLPR moderates the effect of relational peer victimization on depressive symptoms in adolescent girls. *Journal of Child Psychology and Psychiatry, 51*(2), 173–179.

Berntson, G. G., Cacioppo, J. T., & Quigley, K. S. (1991). Autonomic determinism: The modes of autonomic control, the doctrine of autonomic space, and the laws of autonomic constraint. *Psychological Review, 98*(4), 459–487.

Berntson, G. G., Quigley, K. S., & Lozano, D. (2007). Cardiovascular psychophysiology. In J. T. Cacioppo, L. G. Tassinary, & G. G. Berntson (Eds.), *Handbook of psychophysiology* (3rd ed., pp. 182–210). Cambridge University Press.

Berntson, G. G., Quigley, K. S., Norman, G. J., & Lozano, D. L. (2016). Cardiovascular psychophysiology. In J. T. Cacioppo, L. G. Tassinary, & G. G. Berntson (Eds.), *Handbook of psychophysiology* (4th ed., pp. 183–216). Cambridge University Press.

Blackhart, G. C., Eckel, L. A., & Tice, D. M. (2007). Salivary cortisol in response to acute social rejection and acceptance by peers. *Biological Psychology, 75*(3), 267–276.

Bollmer, J. M., Harris, M. J., & Milich, R. (2006). Reactions to bullying and peer victimization: Narratives, physiological arousal, and personality. *Journal of Research in Personality, 40*(5), 803–828.

Boyce, W. T., & Ellis, B. J. (2005). Biological sensitivity to context: I. An evolutionary-developmental theory of the origins and functions of stress reactivity. *Development and Psychopathology, 17*(2), 271–301.

Brendgen, M., Boivin, M., Vitaro, F., Girard, A., Dionne, G., & Pérusse, D. (2008). Gene–environment interaction between peer victimization and child aggression. *Development and Psychopathology, 20*(2), 455–471.

Brendgen, M., Barker, E. D., Boivin, M., Dionne, G., Vitaro, F., Girard, A., Tremblay, R., & Pérusse, D. (2011). Gene–environment processes linking aggression, peer victimization, and the teacher–child relationship. *Child Development, 82*(6), 2021–2036.

Brendgen, M., Ouellet-Morin, I., Lupien, S., Vitaro, F., Dionne, G., & Boivin, M. (2017a). Does cortisol moderate the environmental association between peer victimization and depression symptoms? A genetically informed twin study. *Psychoneuroendocrinology, 84*, 42–50.

Brendgen, M., Ouellet-Morin, I., Lupien, S. J., Vitaro, F., Dionne, G., & Boivin, M. (2017b). Environmental influence of problematic social relationships on adolescents' daily cortisol secretion: A monozygotic twin-difference study. *Psychological Medicine, 47*(3), 460–470.

Breslend, N. L., Shoulberg, E. K., Wagner, C., Murray-Close, D., & Holterman, L. A. (2018). Biosocial interactions between relational victimization and physiological stress reactivity in relation to anxious/depressive symptoms and cognitive biases in adolescent girls. *Merrill-Palmer Quarterly, 64*(1), 41–69.

Cacioppo, J. T., Tassinary, L. G., & Berntson, G. (2007). *Handbook of psychophysiology*. Cambridge University Press.

Calhoun, C. D., Helms, S. W., Heilbron, N., Rudolph, K. D., Hastings, P. D., & Prinstein, M. J. (2014). Relational victimization, friendship, and adolescents' hypothalamic–pituitary–adrenal axis responses to an in vivo social stressor. *Development and Psychopathology, 26*(3), 605–618.

Carney, J. V., Hazler, R. J., Oh, I., Hibel, L. C., & Granger, D. A. (2010). The relations between bullying exposures in middle childhood, anxiety, and adrenocortical activity. *Journal of School Violence, 9*(2), 194–211.

Chen, G., Kong, Y., Deater-Deckard, K., & Zhang, W. (2018). Bullying victimization heightens cortisol response to psychosocial stress in Chinese children. *Journal of Abnormal Child Psychology, 46*(5), 1051–1059.

Copeland, W. E., Wolke, D., Lereya, S. T., Shanahan, L., Worthman, C., & Costello, E. J. (2014). Childhood bullying involvement predicts low-grade systemic inflammation into adulthood. *Proceedings of the National Academy of Sciences, 111*(21), 7570–7575.

de Bruine, M., Giletta, M., Denissen, J. J., Sijtsema, J. J., & Oldehinkel, A. J. (2019). A healthy peer status: Peer preference, not popularity, predicts lower systemic inflammation in adolescence. *Psychoneuroendocrinology, 109*, 104402.

Del Giudice, M., Ellis, B. J., & Shirtcliff, E. A. (2011). The adaptive calibration model of stress responsivity. *Neuroscience & Biobehavioral Reviews, 35*(7), 1562–1592.

Dickerson, S. S., & Kemeny, M. E. (2004). Acute stressors and cortisol responses: A theoretical integration and synthesis of laboratory research. *Psychological Bulletin, 130*(3), 355–391.

DiLalla, L. F., Bersted, K., & John, S. G. (2015). Peer victimization and DRD4 genotype influence problem behaviors in young children. *Journal of Youth and Adolescence, 44*(8), 1478–1493.

Edwards, K. M., Burns, V. E., Reynolds, T., Carroll, D., Drayson, M., & Ring, C. (2006). Acute stress exposure prior to influenza vaccination enhances antibody response in women. *Brain, Behavior, and Immunity, 20*, 159–168.

Eisenberger, N. I. (2012). The pain of social disconnection: Examining the shared neural underpinnings of physical and social pain. *Nature Reviews Neuroscience, 13*(6), 421–434.

Ellis, B. J., Boyce, W. T., Belsky, J., Bakermans-Kranenburg, M. J., & Van IJzendoorn, M. H. (2011). Differential susceptibility to the environment: An evolutionary–neurodevelopmental theory. *Development and Psychopathology, 23*(1), 7–28.

Ellis, B. J., Del Giudice, M., & Shirtcliff, E. A. (2017). The adaptive calibration model of stress responsivity: Concepts, findings, and implications for developmental psychpathology. In T. P. Beauchaine & S. P. Hinshaw (Eds.), *Child and adolescent psychopathology* (3rd ed., pp. 237–276). Wiley & Sons.

Erath, S. A., & Tu, K. M. (2014). Peer stress in preadolescence: Linking physiological and coping responses with social competence. *Journal of Research on Adolescence, 24*(4), 757–771.

Erath, S. A., Tu, K. M., & El-Sheikh, M. (2012). Socially anxious and peer-victimized preadolescents: "Doubly primed" for distress? *Journal of Abnormal Child Psychology, 40*(5), 837–848.

Erath, S. A., Su, S., & Tu, K. M. (2018). Electrodermal reactivity moderates the prospective association between peer victimization and depressive symptoms in early adolescence. *Journal of Clinical Child & Adolescent Psychology, 47*(6), 992–1003.

Gardner, W. L., Pickett, C. L., & Brewer, M. B. (2000). Social exclusion and selective memory: How the need to belong influences memory for social events. *Personality and Social Psychology Bulletin, 26*(4), 486–496.

Giletta, M., Slavich, G. M., Rudolph, K. D., Hastings, P. D., Nock, M. K., & Prinstein, M. J. (2018). Peer victimization predicts heightened inflammatory reactivity to social stress in cognitively vulnerable adolescents. *Journal of Child Psychology and Psychiatry, 59*(2), 129–139.

Graziano, P. A., Keane, S. P., & Calkins, S. D. (2007). Cardiac vagal regulation and early peer status. *Child Development, 78*(1), 264–278.

Gregson, K. D., Tu, K. M., & Erath, S. A. (2014). Sweating under pressure: Skin conductance level reactivity moderates the association between peer victimization and externalizing behavior. *Journal of Child Psychology and Psychiatry, 55*(1), 22–30.

Hamilton, L. D., Newman, M. L., Delville, C. L., & Delville, Y. (2008). Physiological stress response of young adults exposed to bullying during adolescence. *Physiology and Behavior, 95*(5), 617–624.

Iffland, B., Sansen, L. M., Catani, C., & Neuner, F. (2014). The trauma of peer abuse: Effects of relational peer victimization and social anxiety disorder on physiological and affective reactions to social exclusion. *Frontiers in Psychiatry, 5*, 26–34.

Iffland, B., Wiggert, N., Neuner, F., & Blechert, J. (2018). Neutral is negative enough – Peer victimization influences cardiac, facial-muscular and experiential reactions to both negative evaluative and neutral social stimuli. *Biological Psychology, 139*, 152–162.

Irwin, M. R., & Cole, S. W. (2011). Reciprocal regulation of the neural and innate immune systems. *Nature Reviews Immunology, 11*(9), 625–632.

Irwin, M. R., Levin, M. J., Laudenslager, M. L., Olmstead, R., Lucko, A., Lang, N., et al. (2013). Varicella zoster virus-specific immune responses to a herpes zoster vaccine in elderly recipients with major depression and the impact of antidepressant medications. *Clinical Infectious Diseases, 56*, 1085–1093.

Juster, R. P., McEwen, B. S., & Lupien, S. J. (2010). Allostatic load biomarkers of chronic stress and impact on health and cognition. *Neuroscience & Biobehavioral Reviews, 35*(1), 2–16.

Kliewer, W. (2006). Violence exposure and cortisol responses in urban youth. *International Journal of Behavioral Medicine, 13*(2), 109–120.

Kliewer, W. (2016). Victimization and biological stress responses in urban adolescents: Emotion regulation as a moderator. *Journal of Youth and Adolescence, 45*(9), 1812–1823.

Kliewer, W., Dibble, A. E., Goodman, K. L., & Sullivan, T. N. (2012). Physiological correlates of peer victimization and aggression in African American urban adolescents. *Development and Psychopathology, 24*(2), 637–650.

Knack, J. M., Jensen-Campbell, L. A., & Baum, A. (2011). Worse than sticks and stones? Bullying is associated with altered HPA axis functioning and poorer health. *Brain and Cognition, 77*(2), 183–190.

Leary, M. R., Tambor, E. S., Terdal, S. K., & Downs, D. L. (1995). Self-esteem as an interpersonal monitor: The sociometer hypothesis. *Journal of Personality and Social Psychology, 68*(3), 518–530.

Lovallo, W. R., & Buchanan, T. W. (2017). Stress hormones in psychophysiological research: Emotional, behavioral, and cognitive implications. In J. T. Cacioppo, L. G. Tassinary, & G. G. Berntson (Eds.), *Handbook of psychophysiology* (4th ed., pp. 465–494). Cambridge University Press.

Mazzone, A., Camodeca, M., Cardone, D., & Merla, A. (2017). Bullying perpetration and victimization in early adolescence: Physiological response to social exclusion. *International Journal of Developmental Science, 11*(3–4), 121–130.

McLaughlin, K. A., Sheridan, M. A., Tibu, F., Fox, N. A., Zeanah, C. H., & Nelson, C. A. (2015). Causal effects of the early caregiving environment on development of stress response systems in children. *Proceedings of the National Academy of Sciences, 112*(18), 5637–5642.

Mendes, W. B., Major, B., McCoy, S., & Blascovich, J. (2008). How attributional ambiguity shapes physiological and emotional responses to social rejection and acceptance. *Journal of Personality and Social Psychology, 94*(2), 278–291.

Michels, N., Sioen, I., Clays, E., De Buyzere, M., Ahrens, W., Huybrechts, I., Vanaelst, B., & De Henauw, S. (2013). Children's heart rate variability as stress indicator: Association with reported stress and cortisol. *Biological Psychology, 94*(2), 433–440.

Miller, G. E., Chen, E., & Zhou, E. S. (2007). If it goes up, must it come down? Chronic stress and the hypothalamic-pituitary-adrenocortical axis in humans. *Psychological Bulletin, 133*(1), 25.

Monroe, S. M., & Harkness, K. L. (2005). Life stress, the "kindling" hypothesis, and the recurrence of depression: Considerations from a life stress perspective. *Psychological Review, 112*(2), 417–445.

Monti, J. D., Abaied, J. L., & Rudolph, K. D. (2014). Contributions of socialisation of coping to physiological responses to stress. *Australian Journal of Psychology, 66*(2), 102–109.

Murali, R., & Chen, E. (2005). Exposure to violence and cardiovascular and neuroendocrine measures in adolescents. *Annals of Behavioral Medicine, 30*(2), 155–163.

Murphy, M. L., Slavich, G. M., Rohleder, N., & Miller, G. E. (2013). Targeted rejection triggers differential pro-and anti-inflammatory gene expression in adolescents as a function of social status. *Clinical Psychological Science, 1*(1), 30–40.

Murray-Close, D. (2013). Psychophysiology of adolescent peer relations I: Theory and research findings. *Journal of Research on Adolescence, 23*(2), 236–259.

Murray-Close, D., Crick, N. R., Tseng, W. L., Lafko, N., Burrows, C., Pitula, C., & Ralston, P. (2014). Physiological stress reactivity and physical and relational aggression: The moderating roles of victimization, type of stressor, and child gender. *Development and Psychopathology, 26*(3), 589–603.

Newman, M. L. (2014). Here we go again: Bullying history and cardiovascular responses to social exclusion. *Physiology and Behavior, 133*, 76–80.

O'Connor, T. G. (2003). Early experiences and psychological development: Conceptual questions, empirical illustrations, and implications for intervention. *Development and Psychopathology, 15*(3), 671–690.

Ouellet-Morin, I., Danese, A., Bowes, L., Shakoor, S., Ambler, A., Pariante, C. M., et al. (2011). A discordant monozygotic twin design shows blunted cortisol reactivity among bullied children. *Journal of the American Academy of Child and Adolescent Psychiatry, 50*(6), 574–582.

Ouellet-Morin, I., Wong, C. C. Y., Danese, A., Pariante, C. M., Papadopoulos, A. S., Mill, J., & Arseneault, L. (2013). Increased serotonin transporter gene (SERT) DNA methylation is associated with bullying victimization and blunted cortisol response to stress in childhood: A longitudinal study of discordant monozygotic twins. *Psychological Medicine, 43*(9), 1813–1823.

Peters, E., Riksen-Walraven, J. M., Cillessen, A. H., & de Weerth, C. (2011). Peer rejection and HPA activity in middle childhood: Friendship makes a difference. *Child Development, 82*(6), 1906–1920.

Pickett, C. L., Gardner, W. L., & Knowles, M. (2004). Getting a cue: The need to belong and enhanced sensitivity to social cues. *Personality and Social Psychology Bulletin, 30*(9), 1095–1107.

Ponzi, D., Muehlenbein, M. P., Geary, D. C., & Flinn, M. V. (2016). Cortisol, salivary alpha-amylase and children's perceptions of their social networks. *Social Neuroscience, 11*(2), 164–174.

Porges, S. W. (2007). The polyvagal perspective. *Biological Psychology, 74*(2), 116–143.

Rudolph, K. D., & Asher, S. R. (2000). Adaptation and maladaptation in the peer system: Developmental processes and outcomes. In A. J. Sameroff, M. Lewis, & S. M. Miller (Eds.), *Handbook of developmental psychopathology* (2nd ed., pp. 157–175). Springer.

Rudolph, K. D., Troop-Gordon, W., & Granger, D. A. (2010). Peer victimization and aggression: Moderation by individual differences in salivary cortiol and alpha-amylase. *Journal of Abnormal Child Psychology, 38*(6), 843–856.

Rudolph, K. D., Troop-Gordon, W., & Granger, D. A. (2011). Individual differences in biological stress responses moderate the contribution of early peer victimization to subsequent depressive symptoms. *Psychopharmacology, 214*(1), 209–219.

Rudolph, K. D., Lansford, J. E., & Rodkin, P. C. (2016a). Interpersonal theories of psychopathology. In D. Cicchetti (Ed.), *Developmental psychopathology* (3rd ed., pp. 243–311). Wiley.

Rudolph, K. D., Miernicki, M. E., Troop-Gordon, W., Davis, M. M., & Telzer, E. H. (2016b). Adding insult to injury: Neural sensitivity to social exclusion is associated with internalizing symptoms in chronically peer-victimized girls. *Social Cognitive and Affective Neuroscience, 11*(5), 829–842.

Shonkoff, J. P., & Bales, S. N. (2011). Science does not speak for itself: Translating child development research for the public and its policymakers. *Child Development, 82*(1), 17–32.

Slavich, G. M., & Cole, S. W. (2013). The emerging field of human social genomics. *Clinical Psychological Science, 1*(3), 331–348.

Slavich, G. M., & Irwin, M. R. (2014). From stress to inflammation and major depressive disorder: A social signal transduction theory of depression. *Psychological Bulletin, 140*(3), 774–815.

Slavich, G. M., O'Donovan, A., Epel, E. S., & Kemeny, M. E. (2010). Black sheep get the blues: A psychobiological model of social rejection and depression. *Neuroscience and Biobehavioral Reviews, 35*(1), 39–45.

Takizawa, R., Danese, A., Maughan, B., & Arseneault, L. (2015). Bullying victimization in childhood predicts inflammation and obesity at mid-life: A five-decade birth cohort study. *Psychological Medicine, 45*(13), 2705–2715.

Telzer, E. H., Fowler, C. H., Davis, M. M., Miernicki, M. E., & Rudolph, K. D. (2019). Hungry for inclusion: Chronic peer victimization, neural vigilance to peer belonging, and subsequent adjustment. Manuscript submitted for publication.

Tout, K., de Haan, M., Campbell, E. K., & Gunnar, M. R. (1998). Social behavior correlates of cortisol activity in child care: Gender differences and time-of-day effects. *Child Development, 69*(5), 1247–1262.

Vaillancourt, T., Duku, E., Decatanzaro, D., Macmillan, H., Muir, C., & Schmidt, L. A. (2008). Variation in hypothalamic–pituitary–adrenal axis activity among bullied and non-bullied children. *Aggressive Behavior, 34*(3), 294–305.
Vaillancourt, T., Duku, E., Becker, S., Schmidt, L. A., Nicol, J., Muir, C., & MacMillan, H. (2011). Peer victimization, depressive symptoms, and high salivary cortisol predict poorer memory in children. *Brain and Cognition, 77*(2), 191–199.
von Klitzing, K., Perren, S., Klein, A. M., Stadelmann, S., White, L. O., Groeben, M., Holsboer Trachsler, E., Brand, S., & Hatzinger, M. (2012). The interaction of social risk factors and HPA axis dysregulation in predicting emotional symptoms of five-and six-year-old children. *Journal of Psychiatric Research, 46*(3), 290–297.
Williams, S. G. (2017). Mental health issues related to sexual orientation in a high school setting. *The Journal of School Nursing, 33*(5), 383–392.
Williams, S. G., Turner-Henson, A., Davis, S., & Soistmann, H. C. (2017). Relationships among perceived stress, bullying, cortisol, and depressive symptoms in ninth-grade adolescents: A pilot study. *Biological Research for Nursing, 19*(1), 65–70.
Wilson, D. K., Kliewer, W., Teasley, N., Plybon, L., & Sica, D. A. (2002). Violence exposure, catecholamine excretion, and blood pressure nondipping status in African American male versus female adolescents. *Psychosomatic Medicine, 64*(6), 906–915.

School Social Relations, Self-Regulation, and Social Decision-Making

Wendy Troop-Gordon and Stephen A. Erath

Few needs are as fundamental as the need for belongingness (Deci & Ryan, 2008; DeWall et al., 2011). People, including children, have an innate drive to feel connected to others and maintain satisfying relationships. Fulfillment of this need is multifaceted. It requires stable, positive interactions with others; a sense of acceptance by one's family and peers; enduring, emotionally intimate relationships; and a perception that one is valued (Anderman & Leake, 2007; Sullivan, 1953). Thus, fulfillment of belongingness needs can be seen in acceptance by one's peer group, the establishment of high-quality friendships, active engagement with others, popularity within the peer group, and a sense of being respected. However, for some children, interpersonal relationships fail to fulfill this most essential of needs. These children face pervasive peer rejection, friendlessness, or social isolation (Rubin et al., 2005) and, too often, a combination of these.

What happens then when belongingness needs are thwarted? Social psychologists have long documented the biological, cognitive, and affective consequences of transient lapses in belongingness (DeWall et al., 2011). Developmental psychologists have similarly linked poor social relationships to a host of mental and physical health problems. In this chapter, we bridge these two literatures by exploring how pervasive difficulties in forming positive peer relationships at school disrupt regulatory systems and the ability to effectively solve interpersonal problems, placing children's socioemotional health and academic success at risk.

W. Troop-Gordon (✉) · S. A. Erath
Auburn University, Auburn, AL, USA
e-mail: wpg0006@auburn.edu

P. A. C. van Lier, K. Deater-Deckard (eds.), *Biosocial Interplay During Elementary School*, https://doi.org/10.1007/978-3-031-07109-6_7

Self-Regulation Development and Social Decision-Making

Key to competently engaging with others, effectively coping with interpersonal stress, and overcoming obstacles to positive relationship formation is self-regulation. Although definitions vary, we employ the term "self-regulation" here to refer to control over cognition, affect, and behavior in order to satisfy intrapersonal needs and situational demands (Raffaelli et al., 2005). Thus, self-regulation requires a complex set of skills and necessitates the development of neurological systems that support regulatory processes. Self-regulation of cognition, often referred to as "executive function," manifests in flexible and sustained attention to relevant internal and external stimuli, inhibition of attention to distracting information, and inhibitory control over preponderant cognitive responses. Cognitive self-regulation also entails updating working memory and detecting inconsistencies between desired outcomes and current states, allowing for efficient and effective planning (Diamond, 2013). Self-regulation of emotion involves controlling the nature, intensity, and expression of emotions as befits the context (Gross, 2015), while behavioral self-regulation is evidenced by inhibition of impulses, delay of gratification, and effective enactment of response strategies (Raffaelli et al., 2005).

Self-regulation permits control over, and integration of, automatic and reflective response systems. Automatic responses to stress are "relatively unconscious, fast, and efficient," whereas reflective responses are "relatively conscious, deliberate, and effortful" (Evers et al., 2014, p. 44). Examples of automatic responses include physiological reactivity and attentional orienting, and reflective responses include cognitive reappraisals and, of particular importance for this chapter, social decision-making. Both automatic and reflective response systems are requisite for competently achieving interpersonal goals. Self-regulation coordinates the activities of these two systems, including overriding automatic responses when needed to allow for greater reflective responding. Consequently, when self-regulation is immature or disrupted, reactions to interpersonal stress may be disproportionately comprised of automatic responses including heightened physiological reactivity to stress, dysregulated emotional displays, and impulsive, schema-driven behaviors (Rosen et al., 2007; Troop-Gordon et al., 2018).

In contrast, effective self-regulation undergirds reflective decision-making in the face of social challenge. For example, self-regulation is necessary for effectually deploying attention to relevant cues, maintaining motivation toward desired goals, and generating optimal response strategies (Crick & Dodge, 1996; Troop-Gordon & Asher, 2005). Self-regulation also underlies the inhibition of thoughts and behaviors inconsistent with one's objectives, modulation of emotional reactions commensurate with interpersonal goals, and utilization of planning and future-oriented thinking when evaluating response options (Crick & Dodge, 1996; Lemerise & Arsenio, 2000).

Empirical evidence points to normative developmental changes in self-regulation throughout childhood (McClelland et al., 2015). For example, neurobiological advances across childhood and adolescence correspond with increasing inhibitory control, attentional control, error detection, and working memory (Luna et al., 2010, 2015), as well as the utilization of more effective emotion regulation strategies (e.g., cognitive reappraisal, positive refocusing; Ahmed et al., 2015; Mcrae et al., 2012). These cognitive and affective advancements, in turn, underlie improvements in behavioral self-control (Gestsdottir & Lerner, 2008).

Therefore, self-regulation and, in turn, effective social decision-making are developmentally determined skills that vary across children and throughout the school years. Although developmental trajectories of self-regulatory systems are likely multiply-determined, interpersonal stress may have particularly pernicious effects. Thus, we turn to how social stress can undermine self-regulation, resulting in less reflective, competent, social decision-making.

"Under the Microscope": Social Stress and Dysregulation Within Lab-Based Paradigms

Even brief exposure to interpersonal stress can temporarily strain self-regulatory capabilities, at least in the short term. This has been demonstrated using a variety of paradigms in which participants experience brief exclusion or rejection from fictional peers in laboratory settings (e.g., Silk et al., 2011; Williams & Jarvis, 2006). Among adults, such ostracism elicits reactivity in the same brain regions activated by physical pain (Eisenberger et al., 2003), incites dysregulated neural responses (e.g., suppressed pupillary reaction; Sleegers et al., 2017), erodes cognitive self-regulation (Xu et al., 2017), and impairs behavioral self-control (Baumeister et al., 2005). Even among adults who experience normative levels of stress, regaining self-control following ostracism requires effort, as evidenced by heightened activation of brain regions involved in self-regulation (e.g., Vijayakumar et al., 2017), attention to cues of acceptance (Xu et al., 2015), and emotion-focused coping (Boyes & French, 2009).

Lab-based research with children and adolescents shows a similar pattern. Ostracism elicits a heightened neural response, as indicated by pupillary responses (Silk et al., 2011) and brain imaging (Miller et al., 2019). Rejection and exclusion also lead to decrements in working memory (Hawes et al., 2012), heightened emotional reactivity (Miller et al., 2019; Wölfer & Scheithauer, 2013), lower levels of perceived self-control (DeWall et al., 2012), and less persistence at a difficult task (King et al., 2018). Furthermore, like adults, children evidence efforts to regain control after exclusion, as indicated by increased activity in the prefrontal cortex, a major brain region involved in self-regulation (Miller et al., 2019), attentional bias toward acceptance cues (Silk et al., 2011), and selective memory for positive social events (Wölfer & Scheithauer, 2013).

Although compelling, these studies point to only temporary shifts in self-regulation and decision-making that are due to contrived experiences of exclusion and rejection, often by fictitious strangers. Such studies cannot fully explain what happens to children who experience chronic acute levels of peer stress in their daily lives. Do prolonged experiences of interpersonal stress within familiar and meaningful contexts take a sustained toll on children's self-regulatory and decision-making skills? Increasingly, evidence from developmental psychology suggests the answer to this is "yes." For example, in the family domain, interparental conflict leads to diminished self-regulation as reflected in emotional reactivity in threatening contexts and behavioral avoidance (Davies & Cummings, 1998). Upon entry to formal schooling, relationships with peers and teachers take on increasing importance. Consequently, a sense of belonging in school becomes essential for the development of self-regulation and competent social decision-making.

Interpersonal Stress at School: Implications for Self-Regulation and Social Decision-Making

Indeed, accumulating evidence shows that chronic interpersonal stress erodes the self-regulatory processes necessary for careful, reflective decision-making. Here we present a heuristic conceptual framework for understanding the process through which peer stress engenders automatic stress responses and less effective decision-making (see Fig. 1). We propose that chronic peer stress leads to dysregulation of neural and physiological stress response systems. This disrupts intrapsychic processes, including emotion regulation and the adaptive processing of social information. In turn, both emotion dysregulation and poor social information processing can lead to maladaptive coping with stress by (a) eliciting automatic (i.e., involuntary) stress responses and (b) reducing engagement in reflective decision-making. Our goal in this chapter, however, is not to outline evidence of each of these conceptual steps but rather to present evidence that stress in peer relationships at school disrupts self-regulation and effective decision-making.

Peer Stress and Neural/Physiological Response Systems Difficulties in peer relationships at school are hypothesized to "leave a mark" on students' socioemotional well-being by altering neurological reactivity to negative feedback and heightening, or blunting, physiological responses to stress. Will et al. (2016b) have shown that in comparison with adolescents who were accepted by their classmates in elementary school, those who were chronically rejected during elementary school evidence greater reactivity in brain regions involved in processing distress. In addition, when rejected adolescents are deciding whether to share with those who had excluded them, they evidence lower activation of brain regions involved in perspective taking (Will et al., 2016a). Contradicting these findings, however, is evidence of greater responsiveness in brain regions involved in emotion regulation (Asscheman et al., 2019) and forgiveness (Will et al., 2016a) following exclusion among less well-accepted boys. Will et al. (2016a) interpreted these findings as evidence that emotional and behavioral regulation during a challenging interpersonal situation requires

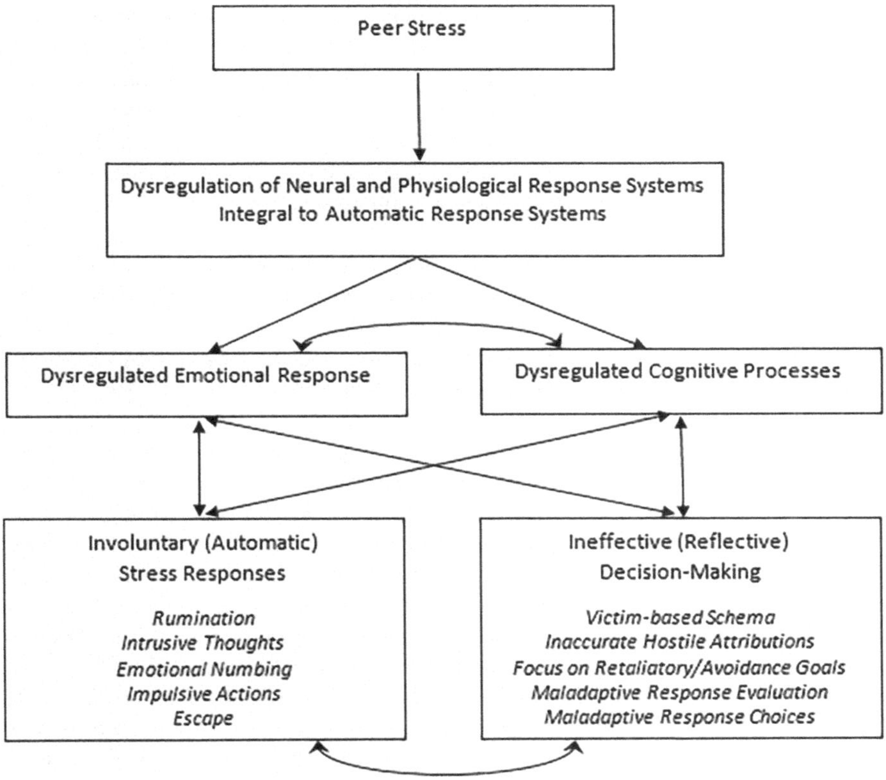

Fig. 1 Heuristic model of how peer stress at school leads to greater engagement of automatic stress responses and impairs reflective decision-making

greater neural regulation among children who have experienced chronic peer stress than among those who have had more positive peer experiences.

Chronic difficulties with peers at school are also linked with dysregulation of the two primary stress response systems: the autonomic nervous system and the hypothalamic-pituitary-adrenal (HPA) axis (see also chapter "The Impact of School Social Experiences on Socioemotional and Behavioral Problems: The Hypothesized Role of DNA Methylation" of this volume). The autonomic nervous system (ANS) prepares the body for "fight or flight" via activation of the sympathetic nervous system and withdrawal of the parasympathetic nervous system. Chronic peer stress is associated with greater sympathetic nervous system activity at rest (Lambe et al., 2019) and in response to a stress-inducing interpersonal task (Gazelle & Druhen, 2009). The HPA axis is a slower responding stress response system that culminates in the production of cortisol. Research linking chronic peer difficulties to cortisol production have been mixed. Although some studies find no association between peer stress and cortisol production (e.g., Rudolph et al., 2010; von Klitzing et al., 2012), others link peer stress to dysregulation in the cortisol awakening response, daily cortisol levels and diurnal slope, and cortisol production in response to stress (e.g., Behnsen et al., 2018; Knack et al., 2011; Ouellet-Morin et al., 2011; Peters et al., 2011; Vaillancourt et al., 2008). Thus, although researchers are still disentangling the association between peer stress and cortisol production (e.g., developmental differences, blunted versus heightened cortisol response; Kliewer et al., 2019), the preponderance of evidence points to peer stress altering automatic, physiological response systems.

Peer Stress and Dysregulated Emotional Response Emotion regulation has long been identified as a critical skill for establishing positive social relationships. The development of emotion regulation, however, is dependent on the quality of children's peer relationships, suggesting that for some students a cycle may emerge in which interpersonal stress and emotion dysregulation are mutually reinforcing.

Studies on dysregulation of the neural/physiological response systems underlying emotion processing and regulation are one avenue through which researchers have identified the damaging effects of chronic peer problems on emotion regulation. A second avenue is through connecting peer experiences directly with poor emotion regulation development. To this end, peer victimization is predictive of lower levels of emotional awareness and control over negative emotions 4 months later (McLaughlin et al., 2009), and chronic loneliness is predictive of less effective emotion regulation in response to social exclusion (Vanhalst et al., 2018). In a study of emotional reactivity to a lab-based interpersonal challenge, Rudolph et al. (2009) found that a history of relational peer victimization was associated with greater negative affect directed at a partner if interactions with that partner were high in conflict. Similarly, Gazelle and Druhen (2009) found that anxious-excluded children evidenced greater distress during an interpersonal rejection paradigm than non-excluded children. Perhaps some of the most compelling evidence, however, comes from studies tracking youth's daily negative experiences and emotional responses to those experiences. Such studies have documented greater negative emotional reactions to peer experiences at school among children who are socially rejected (Morrow et al., 2014) or have no friends to provide support (Reavis et al., 2015).

Peer Stress and Dysregulated Cognitive Processes Contemporary conceptualizations of social competence designate cognitive self-control, or executive functioning, as a core component (Bierman, 2004). Deploying attention flexibly, inhibiting strong impulses, and holding complex social information in mind underlie social behaviors that are responsive to a variety of social cues and circumstances. Suggesting reciprocal connections, developmental models of self-regulation also posit that peer relationships contribute to the emergence and course of executive functions (Farley & Kim-Spoon, 2014). Peer interactions have long been recognized as fertile contexts for executive function development because children are motivated to cooperate and negotiate with equal-status partners to sustain and enhance these relationships (Parker et al., 2006). Consequently, lack of opportunities for constructive peer interaction may thwart the development of executive functions.

Furthermore, according to the strength model of self-control, executive functions are finite resources that can be depleted when overused (Baumeister et al., 2007). Frequent negative peer interactions that require inhibition of strong reactions can tax these regulatory resources, resulting in impaired self-control over time. Indeed, studies have shown that peer problems predict weaker executive functioning. For example, Holmes et al. (2016) examined associations between executive functioning and peer problems (rejection and victimization) from early childhood through middle adolescence. They found that earlier peer problems predicted poorer executive functioning later in childhood. Iyer et al. (2010) similarly reported that peer victimization predicts lower teacher-rated effortful control in middle childhood. Exposure to bullying during early adolescence also contributes to general inattention during early and middle adolescence (Singham et al., 2017), and peer rejection predicts attention deficit hyperactivity disorder symptoms during early to middle childhood (Stenseng et al., 2016).

Peer Stress and Involuntary (Automatic) Stress Responses Dysregulated physiological, emotional, and cognitive response systems culminate in automatic, involuntary reactions to interpersonal stress at the expense of more effortful and reflective responses (Compas et al., 2001). Accordingly, researchers have tried to understand the deleterious effects of negative peer experiences on broader stress response patterns. Troop-Gordon et al. (2017) utilized Compas et al.'s (2001) framework, organizing stress responses into those that encompass effortful (i.e., reflective) engagement, such as problem-solving and cognitive reappraisal, and those that are involuntary (i.e., automatic) and reflect engagement (e.g., rumination) or disengagement from the stressor (e.g., emotional numbing). They documented normative increases in effortful engagement over the middle childhood years and declines in involuntary responses. Importantly, peer victimization deflected these trajectories. Specifically, higher second-grade peer victimization predicted slower growth in effortful engagement coping and weaker declines in involuntary coping.

Examinations of specific stress responses yield a similar pattern. Peer victimization is associated with lower levels of reflective problem-solving (Monti et al., 2017) and more frequent rumination and internalizing coping (Kochenderfer-Ladd & Skinner, 2002; Monti et al., 2017). However, rather than experiencing uncontrollable thoughts and worry, some peer-victimized children respond with greater denial and refusal to think about the stressor (Kochenderfer-Ladd & Skinner, 2002). Evidence that some students who are peer-victimized have difficulty controlling their behaviors, often acting aggressively toward others (Schwartz, 2000), provides further support that chronic peer difficulties can contribute to engagement of automatic response systems in reaction to interpersonal stress.

Peer Stress and Ineffective (Reflective) Decision-Making Even when children who have experienced difficulties forming positive peer relationships engage in problem-solving, their database of past negative social experiences may lead to biases in their social information processing (i.e., attention to social cues, interpretation of events, goal selection, and strategy generation and evaluation) and, in turn, poor behavioral choices (Crick & Dodge, 1996). Thus, in addition to eliciting automatic, dysregulated stress responses, chronic peer stress at school can disrupt efforts at reflective social decision-making.

The cognitive and emotional dysregulation engendered by chronic peer stress may lead to a narrowing of attention and poor attention to available cues (Kaplan et al., 2012). Consequently, rather than drawing upon on available information to make well-informed decisions, children who have experienced chronic peer stress may rely on their social schemas when problem-solving (Crick & Dodge, 1996; Troop-Gordon et al., 2018). Social schemas offer meaning and enhance predictability, allowing children to appraise situations efficiently without relying exclusively on immediately available social cues (Bierman, 2004; Rosen et al., 2007). However, they can also lead to misinterpretations of events and problematic reactions to others. Overreliance on schema-based processing is particularly precarious for children who have experienced chronic peer stress at school, as such children tend to develop schemas of their schoolmates as hostile and of themselves as socially incompetent and likely targets of victimization (Crick & Dodge, 1996; Ladd & Troop-Gordon, 2003; Rosen et al., 2007).

Activation of negative self and peer schemas is theorized to have numerous downstream consequences for the processing of information, including preemptive processing of potentially threatening cues (Rosen et al., 2007; Schippell et al., 2003), such that children actually attend less to potential sources of social threat. Although this may aid in emotion regulation, the consequence may be maladaptive responses to social challenge. For example, Troop-Gordon et al. (2018) found that children who suppress attention when witnessing a child amused at an ambiguously aggressive act tended to show higher levels of aggression. Moreover, this association held only when children held negative peer schema. One interpretation of these findings is that children who hold more negative peer schema quickly construe events as hostile and preemptively suppress further attention to threatening cues, as such cues provide no new information (see also, Horsley et al., 2010; Wilkowski et al., 2007). These results further suggest that inhibition of attention to threatening cues may not sufficiently mitigate negative affect if interpretations of events are derived from hostile social schemas.

Furthermore, studies linking children's peer relationships and social information processing provide direct evidence that severe social difficulties are predictive of making maladaptive attributions for peers' behavior (Crick & Dodge, 1996). Children who are rejected or victimized tend to be quicker to attribute hostility to peers' actions, especially if they themselves are aggressive (Camodeca & Goossens, 2005; De Castro et al., 2002). Other children who have difficult peer relationships are quick to blame their own perceived shortcomings for their interpersonal stress (Graham & Juvonen, 1998).

It follows that children who attribute hostility to other's actions or blame themselves for their peer difficulties pursue quite different goals than children who make more benign self- and other attributions (Salmivalli et al., 2005) and that, in turn, the goals children pursue influence the problem-solving strategies they generate and their evaluation of those strategies (Crick & Dodge, 1996). Using hypothetical conflict vignettes, Troop-Gordon and Asher (2005) found that children rejected by their classmates increasingly focus on goals of retaliation when struggling to resolve a social conflict and are less likely to relinquish instrumental goals (i.e., self-serving goals) than non-rejected children. Corresponding to these goals, rejected children also reported increasingly choosing aggressive tactics during the course of social conflicts and increasing reluctance to acquiesce to the hypothetical peer. Other studies have similarly shown that peer rejection predicts generating fewer solutions to challenging social tasks, providing fewer prosocial solutions, suggesting proportionally more aggressive responses, and selecting more aggressive solutions to problems (e.g., Lansford et al., 2010).

One reason children who experience chronic peer stress at school are quick to adopt goals of retaliation and utilize more aggressive or avoidance strategies is that their previous attempts at using prosocial tactics may have been ineffective. Indeed, a number of strategies that are effective for most children are often ineffectual for children with persistent peer difficulties. For example, Visconti and Troop-Gordon (2010) found that seeking teacher support predicted lower levels of aggression for non-victimized children, but not for victimized children. Kochenderfer-Ladd and Skinner (2002) reported that problem-solving strategies for coping with peer victimization were associated with higher peer preference for non-victimized children, but lower peer preference for victimized children. Thus, engaged, problem-focused coping responses may be less effective for those experiencing chronic peer stress. Over time, repeated failure at using prosocial problem-solving strategies may lead to children quickly resorting to passive or aggressive responses during difficult peer encounters.

There are a number of reasons why effective problem-solving and competent enactment of adaptive strategies may not be effectual for children with chronic peer difficulties. Negative reputations can deter positive peer responses to prosocial behaviors enacted by children who are rejected or victimized (Hymel et al., 1990). Peer victimization may be particularly impervious to engaged, problem-solving efforts, as peer harassment is often perpetrated by more powerful aggressors along with assistants and reinforcers. Consequently, children who are victimized may have limited control to change peer victimization through effective problem-solving (Salmivalli, 2010). Over time, repeated failed attempts to repair peer relationships may lead to social helplessness (i.e., lack of persistence in the face of social challenge; Gazelle & Druhen, 2009; Rudolph et al., 2014).

It is also important to keep in mind that social decision-making that may seem incompetent to others may reflect logical decisions on the part of a child who has experienced chronic peer stress. Due to their negative peer experiences, children who are rejected, victimized, or excluded may choose to protect themselves from further pain and humiliation through preemptive avoidance or aggression. Retaliatory aggression may also function to restore feelings of justice or self-esteem (McDonald, 2019) and relieve anxiety (Visconti & Troop-Gordon, 2010).

Unfortunately, despite the potential short-term protection offered by an orientation toward peers as threatening, decisions to avoid or aggress may contribute to self-fulfilling prophecies that increase the stability of hostile peer treatment over time (Bierman, 2004; McDonald, 2019).

Other School-Based Relationships

Interpersonal stress at school is hardly limited to children's relationships with their peers. The student-teacher relationship is a highly important context for socioemotional development. Yet, many children and youth experience rejection (Mercer & DeRosier, 2008) and high levels of conflict with their teachers (Hamre & Pianta, 2006; Hughes, 2011). Theoretically, stress experienced within the teacher-student relationship should produce the same harmful effects as stress within peer relationships. Unfortunately, researchers have yet to test whether conflict, dismissiveness, and rejection from teachers directly encumbers regulatory processes and erodes children's social decision-making capabilities. Indirect evidence of such effects, however, can be found in research documenting links between poor teacher-student relationships and forms of maladjustment that often arise from regulatory and decision-making deficits. For example, conflictual and low-quality teacher-child relationships are concurrently associated with, and predict over time, poor mental health and behavioral difficulties, including anxiety, depression, and externalizing problems (Jellesma et al., 2015; Mercer & DeRosier, 2008; Pianta & Stuhlman, 2004).

The consequences of poor teacher-child relationships may extend beyond the direct impact of the felt interpersonal stress. Teachers and other school personnel serve as referents with which students evaluate their classmates (Troop-Gordon & Kopp, 2011; White & Kistner, 1992). When teachers express negative affect, frustration, or disappointment with a child, the child's classmates are likely to adopt the teachers' negative view of the child, leading to the child being rejected and victimized by peers (Mercer & DeRosier, 2008; Rudasill et al., 2013; Runions & Shaw, 2013; White & Kistner, 1992). In this manner, stress within the teacher-child relationship may indirectly hinder self-regulation development and social decision-making skills by contributing to stress within peer relationships.

A second means by which teachers may indirectly impact children's self-regulation and decision-making skills is by being a source of emotional and instrumental support when children are experiencing stress in their peer relationships. Children who have established a healthy, warm relationship with a teacher may rely on that teacher for guidance as to how to cope emotionally with peer rejection and victimization and to problem solve means of improving peer experiences. Indeed teacher support can be a powerful buffer against the deleterious effects of peer stress (Spilt et al., 2014; Yeung & Leadbeater, 2010). A reasonable presumption, therefore, is that high-quality teacher-child relations can prevent erosion of self-regulation and social decision-making skills that may otherwise occur as a result of stressful peer experiences.

Implications for School-Based Intervention

Ideally, intervention programs would be implemented that reduce chronic peer stress for all children. Indeed, numerous programs have been developed, and are continuing to be refined, that successfully reduce peer difficulties at school. Some of these programs target specific forms of peer adversity, especially peer victimization (Farrington & Ttofi, 2009). Others promote a general social climate of inclusion and acceptance through cooperative activities and lessons (Mikami et al., 2005). However, these programs are not 100% effective, and there are children whose peer difficulties are too severe to be easily rectified through school-wide curriculum, policy, and intervention. Unfortunately, children who continue to struggle socially despite being in a school where large efforts have been made to promote positive peer relationships often evidence increased socioemotional difficulties (Juvonen & Schacter, 2020).

Thus, school-wide interventions need to be supplemented with efforts targeted at students with extreme and persistent peer difficulties. Such efforts could target not only the behavioral skill deficits that may be contributing to the child's social difficulties but also the self-regulatory deficits and decision-making processes that stem from chronic peer stress. Social skills training programs with behavioral and social-cognitive components (Lochman et al., 1993, 2009) reliably improve targeted social behaviors and cognitions of children who are rejected by their peers. However, generalized and sustained improvements in peer relationships are challenging to achieve for children with significant self-regulatory and peer difficulties. Such improvements may require multifaceted interventions that address additional risk factors in the family and academic domains, and that nest indicated social and emotional skills interventions within universal programs designed to improve the social climate (Bierman & Powers, 2009; Conduct Problems Prevention Research Group, 1999).

In addition to skills training approaches that primarily target social behaviors and cognitions, researchers have recently examined interventions that aim to promote self-regulation through direct practice with self-regulatory skills, such as attentional control, or through enriched social environments and social-emotional learning experiences (Bierman & Torres, 2016). Randomized controlled trials of programs designed to improve teacher-child relationships and classroom management (Raver et al., 2008) as well as social-emotional learning (Greenberg, 2006; Sasser et al., 2017) have revealed positive effects on self-regulatory skills, including executive functioning. These programs create and capitalize on opportunities for self-regulatory skill practice in natural settings with adult scaffolding and support (Bierman & Torres, 2016). The positive effects are consistent with evidence for the developmental significance of positive social experiences as well as the interconnections between regulation of emotions, cognitions, and behaviors.

For students experiencing severe peer stress, careful management of the social environment within the school may provide additional support needed to overcome peer relationship difficulties. For example, separating peer-victimized children from their aggressors (Troop-Gordon & Ladd, 2015), carefully arranging seating assignments (Hamm & Hoffman, 2016), and identifying peers who could act as defenders and offer inclusion in social activities (Salmivalli, 2014) may prevent further peer problems, particularly when children are receiving additional instruction on self-regulation and decision-making. Such efforts, however, require that teachers be highly attuned to the social strengths and problems of their students, as well as the relationship dynamics among all of their students (Hamm & Hoffman, 2016). Efforts to help teachers develop the tools necessary to identify the often very subtle nuances in children's relationships may yield positive results for students struggling to form positive peer relationships, but too often teachers are unprepared to handle their students' peer difficulties (Boulton, 1997; Oldenburg et al., 2016). Thus, an important implication of the work presented here is the need for greater teacher training in identifying children requiring intervention and creating a more positive peer environment in the school for their most socially vulnerable students. Furthermore, counselors are needed who can work individually with children who have self-regulatory or decision-making deficits.

Summary and Conclusion

Although adults often like to believe that the stress children experience with their classmates is temporary and harmless (Troop-Gordon & Gerardy, 2012; Troop-Gordon & Ladd, 2015), belongingness is a fundamental need, and even temporary threats to belonging tap regulatory capabilities. In the context of stable, positive peer relationships, belongingness can be restored quickly, and self-regulatory skills may be advanced as children learn how to effectively respond to stress. Chronic rejection, victimization, friendlessness, and social isolation, however, can serve as impediments to the development of self-regulation and, in turn, effective social decision-making. In this chapter, we presented a conceptual framework with which to understand how peer stress promotes involuntary, dysregulated stress responses and hinders effective, reflective decision-making. Central to this model was dysregulation of the neural and physiological response systems that underlie emotional reactions to stress and cognitive processing of information. Furthermore, as children experience their classmates as aggressive and rejecting, they develop schemas of others as hostile and themselves as socially incompetent. Reliance on these schemas contributes to poor decision-making, as evidenced by misattributions for events, avoidant or retaliatory goals, and problematic response choices.

As this chapter shows, there is substantial research connecting peer difficulties at school with poor self-regulation and decision-making. From a theoretical perspective, research is needed to test the conceptual model presented here. In particular, more research should be devoted to how dysregulated emotional responses and poorly developed executive functions are associated concurrently and longitudinally with heightened involuntary responses to stress and ineffective reflective decision-making. Understanding the most critical self-regulatory processes that allow some children to uncouple decision-making and behavioral responses from dysregulated cognitive and emotional responses would inform intervention. Established lab-based paradigms (e.g., cyberball, chat room tasks) allow for a microgenetic analysis of stress-neurobiology-regulation-response pathways, while longitudinal research can elucidate the developmental unfolding of lasting impact of chronic peer stress on stable within-individual multilevel changes in responses to stress and decision-making. Such basic developmental research should be complemented with intervention studies that test change in responses to stress, potentially producing stronger theoretical models and empirical evidence.

Furthermore, peer stress is predictive of long-term socioemotional and academic problems (Rubin et al., 2006). Understanding the extent to which disruptions to self-regulation and social decision-making account for those links would allow school personnel and clinicians to address underlying mechanisms before severe problems develop. However, even without such knowledge, the extant findings unarguably serve as a call for greater attention to students' feelings of belongingness at school, including systemic efforts to identify those students at greatest risk for peer rejection and isolation and coordinated efforts to help those students foster positive relationships.

References

Ahmed, S. P., Bittencourt-Hewitt, A., & Sebastian, C. L. (2015). Neurocognitive bases of emotion regulation development in adolescence. *Developmental Cognitive Neuroscience, 15*, 11–25.

Anderman, L. H., & Leake, V. S. (2007). The interface of school and family in meeting the belonging needs of early adolescence. In S. Mertens, V. Anfara, & M. Caskey (Eds.), *Handbook of research in middle level education, Vol. 6. The young adolescent and middle school* (pp. 163–182). Information Age Publishing.

Asscheman, J. S., Koot, S., Ma, I., Buil, J. M., Krabbendam, L., Cillessen, A. H., & van Lier, P. A. (2019). Heightened neural sensitivity to social exclusion in boys with a history of low peer preference during primary school. *Developmental Cognitive Neuroscience, 38*, 100673.

Baumeister, R. F., DeWall, C. N., Ciarocco, N. J., & Twenge, J. M. (2005). Social exclusion impairs self-regulation. *Journal of Personality and Social Psychology, 88*, 589–604.

Baumeister, R. F., Vohs, K. D., & Tice, D. M. (2007). The strength model of self-control. *Current Directions in Psychological Science, 16*, 351–355.

Behnsen, P., Buil, M., Koot, S., Huizink, A., & van Lier, P. (2018). Classroom social experiences in early elementary school relate to diurnal cortisol levels. *Psychoneuroendocrinology, 87*, 1–8.

Bierman, K. L. (2004). *Peer rejection: Developmental processes and intervention strategies*. Guilford Press.

Bierman, K. L., & Powers, C. J. (2009). Social skills training to improve peer relations. In K. H. Rubin, W. M. Bukowski, & B. Laursen (Eds.), *Handbook of peer interactions, relationships, and groups* (pp. 63–81). The Guilford Press.

Bierman, K. L., & Torres, M. (2016). Promoting the development of executive functions through early education and prevention programs. In J. A. Griffin, L. S. Freund, & P. McCardle (Eds.), *Executive function in preschool age children: Integrating measurement, neurodevelopment and translational research* (pp. 299–326). American Psychological Association.

Boulton, M. J. (1997). Teachers' views on bullying: Definitions, attitudes and ability to cope. *British Journal of Educational Psychology, 67*, 223–233.

Boyes, M. E., & French, D. J. (2009). Having a Cyberball: Using a ball-throwing game as an experimental social stressor to examine the relationship between neuroticism and coping. *Personality and Individual Differences, 47*, 396–401.

Camodeca, M., & Goossens, F. A. (2005). Aggression, social cognitions, anger and sadness in bullies and victims. *Journal of Child Psychology and Psychiatry, 46*, 186–197.

Compas, B. E., Connor-Smith, J. K., Saltzman, H., Thomsen, A. H., & Wadsworth, M. E. (2001). Coping with stress during childhood and adolescence: Problems, progress, and potential in theory and research. *Psychological Bulletin, 127*, 87–127.

Conduct Problems Prevention Research Group. (1999). Initial impact of the fast track prevention trial for conduct problems: I. the high-risk sample. *Journal of Consulting and Clinical Psychology, 67*, 631–647.

Crick, N. R., & Dodge, K. A. (1996). Social information-processing mechanisms in children's social adjustment. *Psychological Bulletin, 115*, 74–101.

Davies, P. T., & Cummings, E. M. (1998). Exploring children's emotional security as a mediator of the link between marital relations and child adjustment. *Child Development, 69*, 124–139.

De Castro, B. O., Veerman, J. W., Koops, W., Bosch, J. D., & Monshouwer, H. J. (2002). Hostile attribution of intent and aggressive behavior: A meta-analysis. *Child Development, 73*, 916–934.

Deci, E. L., & Ryan, R. M. (2008). Self-determination theory: A macrotheory of human motivation, development, and health. *Canadian Psychology, 49*, 182–185.

DeWall, C. N., Deckman, T., Pond, R. S., Jr., & Bonser, I. (2011). Belongingness as a core personality trait: How social exclusion influences social functioning and personality expression. *Journal of Personality, 79*, 1281–1314.

DeWall, C. N., Gilman, R., Sharif, V., Carboni, I., & Rice, K. G. (2012). Left out, sluggardly, and blue: Low self-control mediates the relationship between ostracism and depression. *Personality and Individual Differences, 53*, 832–837.

Diamond, A. (2013). Executive functions. *Annual Review of Psychology, 64*, 135–168.

Eisenberger, N. I., Lieberman, M. D., & Williams, K. D. (2003). Does rejection hurt? An fMRI study of social exclusion. *Science, 302*, 290–292.

Evers, C., Hopp, H., Gross, J. J., Fischer, A. H., Manstead, A. S., & Mauss, I. B. (2014). Emotion response coherence: A dual-process perspective. *Biological Psychology, 98*, 43–49.

Farley, J. P., & Kim-Spoon, J. (2014). The development of adolescent self-regulation: Reviewing the role of parent, peer, friend, and romantic relationships. *Journal of Adolescence, 37*, 433–440.

Farrington, D. P., & Ttofi, M. M. (2009). School-based programs to reduce bullying and victimization. *Campbell Systematic Reviews, 5*, i–148.

Gazelle, H., & Druhen, M. J. (2009). Anxious solitude and peer exclusion predict social helplessness, upset affect, and vagal regulation in response to behavioral rejection by a friend. *Developmental Psychology, 45*, 1077–1096.

Gestsdottir, S., & Lerner, R. M. (2008). Positive development in adolescence: The development and role of intentional self-regulation. *Human Development, 51*, 202–224.

Graham, S., & Juvonen, J. (1998). Self-blame and peer victimization in middle school: An attributional analysis. *Developmental Psychology, 34*(3), 587–599. https://doi.org/10.1037/0012-1649.34.3.587

Greenberg, M. T. (2006). Promoting resilience in children and youth: Preventive interventions and their interface with neuroscience. *Annals of the New York Academy of Sciences, 1094*, 139–150.

Gross, J. J. (2015). Emotion regulation: Current status and future prospects. *Psychological Inquiry, 26*, 1–26.

Hamm, J. V., & Hoffman, A. S. (2016). Teachers' influence on students' peer relationships and peer ecologies. In K. R. Wentzel & G. B. Ramani (Eds.), *Handbook of social influences in school contexts: Social-emotional, motivation, and cognitive outcomes* (pp. 208–229). Routledge.

Hamre, B. K., & Pianta, R. C. (2006). Student-teacher relationships. In G. G. Bear & K. M. Minke (Eds.), *Children's needs III: Development, prevention, and intervention* (pp. 59–71). National Association of School Psychologists.

Hawes, D. J., Zadro, L., Fink, E., Richardson, R., O'Moore, K., Griffiths, B., et al. (2012). The effects of peer ostracism on children's cognitive processes. *European Journal of Developmental Psychology, 9*, 599–613.

Holmes, C. J., Kim-Spoon, J., & Deater-Deckard, K. (2016). Linking executive function and peer problems from early childhood through middle adolescence. *Journal of Abnormal Child Psychology, 44*, 31–42.

Horsley, T. A., de Castro, B. O., & Van der Schoot, M. (2010). In the eye of the beholder: Eye-tracking assessment of social information processing in aggressive behavior. *Journal of Abnormal Child Psychology, 38*, 587–599.

Hughes, J. N. (2011). Longitudinal effects of teacher and student perceptions of teacher-student relationship qualities on academic adjustment. *The Elementary School Journal, 112*, 38–60.

Hymel, S., Wagner, E., & Butler, L. J. (1990). Reputational bias: View from the peer group. In S. R. Asher & J. D. Coie (Eds.), *Peer rejection in childhood* (pp. 156–186). Cambridge University Press.

Iyer, R. V., Kochenderfer-Ladd, B., Eisenberg, N., & Thompson, M. (2010). Peer victimization and effortful control: Relations to school engagement and academic achievement. *Merrill-Palmer Quarterly, 56*, 361–387.

Jellesma, F. C., Zee, M., & Koomen, H. M. (2015). Children's perceptions of the relationship with the teacher: Associations with appraisals and internalizing problems in middle childhood. *Journal of Applied Developmental Psychology, 36*, 30–38.

Juvonen, J., & Schacter, H. L. (2020). When low rates of bullying increase risks for those who are bullied: the safe school paradox. *JAMA Pediatrics, 174*(4), 317–318. https://doi.org/10.1001/jamapediatrics.2019.5888

Kaplan, R. L., Van Damme, I., & Levine, L. J. (2012). Motivation matters: Differing effects of pre-goal and post-goal emotions on attention and memory. *Frontiers in Psychology, 3*, 404.

King, K. M., McLaughlin, K. A., Silk, J., & Monahan, K. C. (2018). Peer effects on self-regulation in adolescence depend on the nature and quality of the peer interaction. *Development and Psychopathology, 30*(4), 1389–1401. https://doi.org/10.1017/S0954579417001560

Kliewer, W., Sosnowski, D. W., Noh, H., McGuire, K., & Wright, A. W. (2019). Peer victimization and cortisol production in children and adolescents: A systematic review. *Journal of Applied Biobehavioral Research, 24*(4), e12172. https://doi.org/10.1111/jabr.12172

Knack, J. M., Jensen-Campbell, L. A., & Baum, A. (2011). Worse than sticks and stones? Bullying is associated with altered HPA axis functioning and poorer health. *Brain and Cognition, 77*, 183–190.

Kochenderfer-Ladd, B., & Skinner, K. (2002). Children's coping strategies: Moderators of the effects of peer victimization? *Developmental Psychology, 38*, 267–278.

Ladd, G. W., & Troop-Gordon, W. (2003). The role of chronic peer difficulties in the development of children's psychological adjustment problems. *Child Development, 74*, 1344–1367.

Lambe, L. J., Craig, W. M., & Hollenstein, T. (2019). Blunted physiological stress reactivity among youth with a history of bullying and victimization: Links to depressive symptoms. *Journal of Abnormal Child Psychology, 47*(12), 1981–1993. https://doi.org/10.1007/s10802-019-00565-y

Lansford, J. E., Malone, P. S., Dodge, K. A., Pettit, G. S., & Bates, J. E. (2010). Developmental cascades of peer rejection, social information processing biases, and aggression during middle childhood. *Development and Psychopathology, 22*, 593–602.

Lemerise, E. A., & Arsenio, W. F. (2000). An integrated model of emotion processes and cognition in social information processing. *Child Development, 71*, 107–118.

Lochman, J. E., Coie, J. D., Underwood, M. K., & Terry, R. (1993). Effectiveness of a social relations intervention program for aggressive and nonaggressive, rejected children. *Journal of Consulting and Clinical Psychology, 61*, 1053–1058.

Lochman, J. E., Boxmeyer, C., Powell, N., Qu, L., Wells, K., & Windle, M. (2009). Dissemination of the coping power program: Importance of intensity of counselor training. *Journal of Consulting and Clinical Psychology, 77*, 397–409.

Luna, B., Padmanabhan, A., & O'Hearn, K. (2010). What has fMRI told us about the development of cognitive control through adolescence? *Brain and Cognition, 72*, 101–113.

Luna, B., Marek, S., Larsen, B., Tervo-Clemmens, B., & Chahal, R. (2015). An integrative model of the maturation of cognitive control. *Annual Review of Neuroscience, 38*, 151–170.

McClelland, M. M., Geldhof, J. G., Cameron, C. E., & Wanless, S. B. (2015). Development and self-regulation. In W. F. Overton, P. C. M. Molenaar, & R. M. Lerner (Eds.), *Handbook of child psychology and developmental science* (Vol. 1, pp. 523–565). Wiley.

McDonald, K. L. (2019). Revenge goals. In M. H. Jones (Ed.), *Social goals in the classroom: Findings on student motivation and peer relations* (pp. 111–130). Routledge.

McLaughlin, K. A., Hatzenbuehler, M. L., & Hilt, L. M. (2009). Emotion dysregulation as a mechanism linking peer victimization to internalizing symptoms in adolescents. *Journal of Consulting and Clinical Psychology, 77*, 894–904.

McRae, K., Gross, J. J., Weber, J., Robertson, E. R., Sokol-Hessner, P., Ray, R. D., et al. (2012). The development of emotion regulation: An fMRI study of cognitive reappraisal in children, adolescents and young adults. *Social Cognitive and Affective Neuroscience, 7*, 11–22.

Mercer, S. H., & DeRosier, M. E. (2008). Teacher preference, peer rejection, and student aggression: A prospective study of transactional influence and independent contributions to emotional adjustment and grades. *Journal of School Psychology, 46*, 661–685.

Mikami, A. Y., Boucher, M. A., & Humphreys, K. (2005). Prevention of peer rejection through a classroom-level intervention in middle school. *Journal of Primary Prevention, 26*, 5–23.

Miller, A. B., Prinstein, M. J., Munier, E., Machlin, L. S., & Sheridan, M. A. (2019). Emotion reactivity and regulation in adolescent girls following an interpersonal rejection. *Journal of Cognitive Neuroscience, 31*, 249–261.

Monti, J. D., Rudolph, K. D., & Miernicki, M. E. (2017). Rumination about social stress mediates the association between peer victimization and depressive symptoms during middle childhood. *Journal of Applied Developmental Psychology, 48*, 25–32.

Morrow, M. T., Hubbard, J. A., Barhight, L. J., & Thomson, A. K. (2014). Fifth-grade children's daily experiences of peer victimization and negative emotions: Moderating effects of sex and peer rejection. *Journal of Abnormal Child Psychology, 42*, 1089–1102.

Oldenburg, B., Bosman, R., & Veenstra, R. (2016). Are elementary school teachers prepared to tackle bullying? A pilot study. *School Psychology International, 37*, 64–72.

Ouellet-Morin, I., Danese, A., Bowes, L., Shakoor, S., Ambler, A., Pariante, C. M., et al. (2011). A discordant monozygotic twin design shows blunted cortisol reactivity among bullied children. *Journal of the American Academy of Child & Adolescent Psychiatry, 50*, 574–582.

Parker, J. G., Rubin, K. H., Erath, S. A., Wojslawowicz, J. C., & Buskirk, A. A. (2006). Peer relationships, child development, and adjustment: A developmental psychopathology perspective. In D. Cicchetti & D. J. Cohen (Eds.), *Developmental psychopathology: Theory and method* (Vol. 1, pp. 419–493). Wiley.

Peters, E., Riksen-Walraven, J. M., Cillessen, A. H., & de Weerth, C. (2011). Peer rejection and HPA activity in middle childhood: Friendship makes a difference. *Child Development, 82*, 1906–1920.

Pianta, R. C., & Stuhlman, M. W. (2004). Teacher-child relationships and children's success in the first years of school. *School Psychology Review, 33*, 444–458.

Raffaelli, M., Crockett, L. J., & Shen, Y.-L. (2005). Developmental stability and change in self-regulation from childhood to adolescence. *The Journal of Genetic Psychology, 166*, 54–76.

Raver, C. C., Jones, S. M., Li-Grining, C. P., Metzger, M., Champion, K. M., & Sardin, L. (2008). Improving preschool classroom processes: Preliminary findings from a randomized trial implemented in head start settings. *Early Childhood Research Quarterly, 23*, 10–26.

Reavis, R. D., Donohue, L. J., & Upchurch, M. C. (2015). Friendship, negative peer experiences, and daily positive and negative mood. *Social Development, 24*, 833–851.

Rosen, P. J., Milich, R., & Harris, M. J. (2007). Victims of their own cognitions: Implicit social cognitions, emotional distress, and peer victimization. *Journal of Applied Developmental Psychology, 28*, 211–226.

Rubin, K. H., Coplan, R. J., Chen, X., Buskirk, A., & Wojslawowicz, J. (2005). Peer relationships in childhood. In M. Bornstein & M. Lamb (Eds.), *Developmental science: An advanced textbook* (pp. 469–512). Erlbaum.

Rubin, K. H., Bukowski, W. M., & Parker, J. G. (2006). Peer interactions, relationships, and groups. In N. Eisenberg, W. Damon, & R. M. Lerner (Eds.), *Handbook of child psychology: Vol. 3, Social, emotional, and personality development* (6th ed., pp. 571–645). Wiley.

Rudasill, K. M., Niehaus, K., Buhs, E., & White, J. M. (2013). Temperament in early childhood and peer interactions in third grade: The role of teacher–child relationships in early elementary grades. *Journal of School Psychology, 51*(6), 701–716. https://doi.org/10.1016/j.jsp.2013.08.002

Rudolph, K. D., Troop-Gordon, W., & Flynn, M. (2009). Relational victimization predicts children's social-cognitive and self-regulatory responses in a challenging peer context. *Developmental Psychology, 45*, 1444–1454.

Rudolph, K. D., Troop-Gordon, W., & Granger, D. A. (2010). Peer victimization and aggression: Moderation by individual differences in salivary cortiol and alpha-amylase. *Journal of Abnormal Child Psychology, 38*, 843–856.

Rudolph, K. D., Troop-Gordon, W., Monti, J. D., & Miernicki, M. E. (2014). Moving against and away from the world: The adolescent legacy of peer victimization. *Development and Psychopathology, 26*, 721–734.

Runions, K. C., & Shaw, T. (2013). Teacher–child relationship, child withdrawal and aggression in the development of peer victimization. *Journal of Applied Developmental Psychology, 34*, 319–327.

Salmivalli, C. (2010). Bullying and the peer group: A review. *Aggression and Violent Behavior, 15*, 112–120.

Salmivalli, C. (2014). Participant roles in bullying: How can peer bystanders be utilized in interventions? *Theory Into Practice, 53*, 286–292.

Salmivalli, C., Ojanen, T., Haanpää, J., & Peets, K. (2005). "I'm OK but you're not" and other peer-relational schemas: Explaining individual differences in children's social goals. *Developmental Psychology, 41*, 363–375.

Sasser, T. R., Bierman, K. L., Heinrichs, B., & Nix, R. L. (2017). Preschool intervention can promote sustained growth in the executive-function skills of children exhibiting early deficits. *Psychological Science, 28*, 1719–1730.

Schippell, P. L., Vasey, M. W., Cravens-Brown, L. M., & Bretveld, R. A. (2003). Suppressed attention to rejection, ridicule, and failure cues: A unique correlate of reactive but not proactive aggression in youth. *Journal of Clinical Child and Adolescent Psychology, 32*, 40–55.

Schwartz, D. (2000). Subtypes of victims and aggressors in children's peer groups. *Journal of Abnormal Child Psychology, 28*, 181–192.

Silk, J. S., Stroud, L. R., Siegle, G. J., Dahl, R. E., Lee, K. H., & Nelson, E. E. (2011). Peer acceptance and rejection through the eyes of youth: Pupillary, eyetracking and ecological data from the Chatroom Interact task. *Social Cognitive and Affective Neuroscience, 7*, 93–105.

Singham, T., Viding, E., Schoeler, T., Arseneault, L., Ronald, A., Cecil, C. M., et al. (2017). Concurrent and longitudinal contribution of exposure to bullying in childhood to mental health: The role of vulnerability and resilience. *JAMA Psychiatry, 74*, 1112–1119.

Sleegers, W. W., Proulx, T., & Van Beest, I. (2017). The social pain of Cyberball: Decreased pupillary reactivity to exclusion cues. *Journal of Experimental Social Psychology, 69*, 187–200.

Spilt, J. L., Van Lier, P. A., Leflot, G., Onghena, P., & Colpin, H. (2014). Children's social self-concept and internalizing problems: The influence of peers and teachers. *Child Development, 85*(3), 1248–1256.

Stenseng, F., Belsky, J., Skalicka, V., & Wichstrøm, L. (2016). Peer rejection and attention deficit hyperactivity disorder symptoms: Reciprocal relations through ages 4, 6, and 8. *Child Development, 87*, 365–373.

Sullivan, H. S. (1953). *The interpersonal theory of psychiatry*. Norton.

Troop-Gordon, W., & Asher, S. R. (2005). Modifications in children's goals when encountering obstacles to conflict resolution. *Child Development, 76*, 568–582.

Troop-Gordon, W., & Gerardy, H. (2012). Parents' beliefs about peer victimization and children's socio-emotional development. *Journal of Applied Developmental Psychology, 33*, 40–52.

Troop-Gordon, W., & Kopp, J. (2011). Teacher-child relationship quality and children's peer victimization and aggressive behavior in late childhood. *Social Development, 20*, 536–561.

Troop-Gordon, W., & Ladd, G. W. (2015). Teachers' victimization-related beliefs and strategies: Associations with students' aggressive behavior and peer victimization. *Journal of Abnormal Child Psychology, 43*, 45–60.

Troop-Gordon, W., Sugimura, N., & Rudolph, K. D. (2017). Responses to interpersonal stress: Normative changes across childhood and the impact of peer victimization. *Child Development, 88*, 640–657.

Troop-Gordon, W., Gordon, R. D., Vogel-Ciernia, L., Ewing Lee, E., & Visconti, K. J. (2018). Visual attention to dynamic scenes of ambiguous provocation and children's aggressive behavior. *Journal of Clinical Child & Adolescent Psychology, 47*, 925–940.

Vaillancourt, T., Duku, E., Decatanzaro, D., Macmillan, H., Muir, C., & Schmidt, L. A. (2008). Variation in hypothalamic–pituitary–adrenal axis activity among bullied and non-bullied children. *Aggressive Behavior: Official Journal of the International Society for Research on Aggression, 34*, 294–305.

Vanhalst, J., Luyckx, K., Van Petegem, S., & Soenens, B. (2018). The detrimental effects of adolescents' chronic loneliness on motivation and emotion regulation in social situations. *Journal of Youth and Adolescence, 47*, 162–176.

Vijayakumar, N., Cheng, T. W., & Pfeifer, J. H. (2017). Neural correlates of social exclusion across ages: A coordinate-based meta-analysis of functional MRI studies. *NeuroImage, 153*, 359–368.

Visconti, K. J., & Troop-Gordon, W. (2010). Prospective relations between children's responses to peer victimization and their socioemotional adjustment. *Journal of Applied Developmental Psychology, 31*, 261–272.

von Klitzing, K., Perren, S., Klein, A. M., Stadelmann, S., White, L. O., Groeben, M., et al. (2012). The interaction of social risk factors and HPA axis dysregulation in predicting emotional symptoms of five-and six-year-old children. *Journal of Psychiatric Research, 46*, 290–297.

White, K. J., & Kistner, J. (1992). The influence of teacher feedback on young children's peer preferences and perceptions. *Developmental Psychology, 28*, 933–940.

Wilkowski, B. M., Robinson, M. D., Gordon, R. D., & Troop-Gordon, W. (2007). Tracking the evil eye: Trait anger and selective attention within ambiguously hostile scenes. *Journal of Research in Personality, 41*, 650–666.

Will, G. J., Crone, E. A., Van Lier, P. A., & Güroğlu, B. (2016a). Neural correlates of retaliatory and prosocial reactions to social exclusion: Associations with chronic peer rejection. *Developmental Cognitive Neuroscience, 19*, 288–297.

Will, G. J., van Lier, P. A., Crone, E. A., & Güroğlu, B. (2016b). Chronic childhood peer rejection is associated with heightened neural responses to social exclusion during adolescence. *Journal of Abnormal Child Psychology, 44*, 43–55.

Williams, K. D., & Jarvis, B. (2006). Cyberball: A program for use in research on interpersonal ostracism and acceptance. *Behavior Research Methods, 38*, 174–180.

Wölfer, R., & Scheithauer, H. (2013). Ostracism in childhood and adolescence: Emotional, cognitive, and behavioral effects of social exclusion. *Social Influence, 8*, 217–236.

Xu, M., Li, Z., Zhang, J., Sun, L., Fan, L., Zeng, Q., & Yang, D. (2015). Social exclusion influences attentional bias to social information. *Asian Journal of Social Psychology, 18*, 199–208.

Xu, M., Qiao, L., Qi, S., Li, Z., Diao, L., Fan, L., et al. (2017). Social exclusion weakens storage capacity and attentional filtering ability in visual working memory. *Social Cognitive and Affective Neuroscience, 13*, 92–101.

Yeung, R., & Leadbeater, B. (2010). Adults make a difference: The protective effects of parent and teacher emotional support on emotional and behavioral problems of peer-victimized adolescents. *Journal of Community Psychology, 38*, 80–98.

School Social Relationships and Brain Functioning

J. Susanne Asscheman and Kirby Deater-Deckard

Decades of research has shown that peer relationships during elementary school are highly influential for children's social and emotional functioning and psychological adjustment (see Chaps. 1 and 2 of this book). With the transition to elementary school and as they develop across middle childhood, children spend more time with their peers and less time with family members. The quantity and quality of social experiences with peers are likely to interact with the developing brain to shape how children respond in peer as well as other social interactions and contexts. With the advancement in neuroimaging techniques, it has become easier to measure brain responses to such environmental input across development. This is an important advancement, because studying how the brains of children respond to social environmental input elucidates the underlying mechanisms by which peer experiences may shape development. In addition to informing the basic scientific literature on child development, this information about the developing nervous system can inform assessment, diagnosis, prevention, and intervention.

Most of the prior studies assessing neural responses to social experiences with peers during development have focused on adolescence, in part because this period of development is characterized by heightened sensitivity to peer acceptance and rejection compared to childhood and adulthood (Guyer & Jarcho, 2018). However, early to middle childhood (spanning the time between infancy and puberty) is also an important period for social development. Compared to early childhood, once children begin schooling (in most countries, by age 5 years), they spend significant amounts of time interacting with peers in formal (e.g., school) and informal (e.g.,

J. S. Asscheman (✉)
Clinical Developmental Psychology, Faculty of Behavioral and Movement Sciences, Vrije Universiteit Amsterdam, Amsterdam, The Netherlands

K. Deater-Deckard
Department of Psychological and Brain Sciences, University of Massachusetts, Amherst, MA, USA
e-mail: kdeaterdeck@umass.edu

P. A. C. van Lier, K. Deater-Deckard (eds.), *Biosocial Interplay During Elementary School*, https://doi.org/10.1007/978-3-031-07109-6_8

sports, neighborhood) social settings. Interactions during this period are mostly focused on peer play in dyads or in larger peer groups (Coplan & Arbeau, 2009). This middle childhood period, as opposed to adolescence, has been relatively understudied in the developmental social neuroscience literature—although recently, there has been an increase in research on brain development in middle childhood. From a developmental neuroscience perspective, middle childhood is just as important to study as adolescence and early childhood, because many brain structures and processes that underlie social behavior undergo major changes across middle childhood and attain a mature form before the start of puberty (Giedd et al., 1999; Gogtay et al., 2004).

Furthermore, individual differences in neural sensitivity to social information during adolescence have their origins earlier in development. Thus, social experiences during middle childhood shape brain development in ways that influence individual differences in social behavior during adolescence (Andersen, 2003; Schriber & Guyer, 2016). In addition, the processes of interest are not limited to social contexts and behaviors; social experiences in childhood can have profound effects on other developmental outcomes, such as academic functioning and achievement. For example, a child who has many negative peer experiences (e.g., victimization, rejection) during childhood may be less motivated to perform well in elementary school or may even drop out of school at some point, which will continue to influence outcomes later in the lifespan (e.g., job opportunities, health literacy).

The goal of the current chapter is to describe the advancement in the field of developmental social neuroscience during childhood with a specific focus on peer experiences. We begin by presenting a theoretical framework designating how peer social experiences, and more specifically social stressors, may impact brain function during development. We then provide an overview of empirical studies assessing brain responses to peer social experiences during childhood. To provide a coherent and detailed overview, we have concentrated on neuroimaging studies describing any sort of social interaction with peers. These studies of brain responses include experimental manipulation of participants receiving feedback and experiencing rejection, as well as naturalistic correlational studies investigating potential influences of prior real-life peer experiences. We included papers that span the preschool and elementary school phases of development, with an age range between 4 and 12 years old. Additionally, to be comprehensive, we included studies that used electroencephalography (EEG) or functional magnetic resonance imaging (fMRI); EEG is more commonly used than MRI with younger children. We conclude by providing directions for future research on peer social experiences and brain development during childhood.

Theoretical Foundations

Social connectedness is fundamental for human survival and developmental thriving. Like other social species, humans are highly motivated to form and maintain social bonds, so they actively search for and engage in social interactions with others. For instance, adults who have experienced exclusion show behaviors that support affiliations with others such as higher levels of prosocial behavior or will adjust their behavior in accordance with others to fit in (i.e., tend and befriend; Taylor, 2011). The socially embedded context for humans has corresponded with the evolutionary emergence of several networks throughout the brain that seem to be involved in detecting and interpreting social information and engaging in sustained social interactions and enduring relationships (Kilford et al., 2016; Nelson et al., 2005; Silston et al., 2018; Stanley & Adolphs, 2013).

Specifically, the *detection node* processes perceptual information to determine whether a stimulus is social (e.g., human or animate) versus nonsocial. The detection node includes the inferior occipital cortex, inferior temporal lobe, intraparietal sulcus, fusiform gyrus (also known as the fusiform face area that is involved in processing facial information), and the superior temporal sulcus. The *affective node* processes detected social information to compute its affective valence and salience. Regions include the amygdala, ventral striatum, septum, bed nucleus of the stria, hypothalamus, orbitofrontal cortex, anterior cingulate cortex, and anterior insula. The *cognitive-regulatory node* is implicated in mentalizing activities (e.g., inferring cause and meaning from social stimuli), inhibiting prepotent responses, and sustaining goal-directed behavior. Mentalizing processes have been associated with activation in the paracingulate cortex and dorsomedial prefrontal cortex; inhibitory control has been associated with activation in the ventral prefrontal cortex; and goal-directed behavior appears to be supported through the dorsal and ventral prefrontal cortex.

Given the importance of social bonds to human development and functioning, being rejected or excluded by others is typically experienced as moderately to highly aversive and causes a stress response characterized by the upregulation of stress hormones (e.g., cortisol) in the body (Dickerson & Zoccola, 2013). At the neural level of analysis, experiences of rejection or exclusion activate the anterior insula, anterior cingulate cortex, posterior cingulate cortex, temporal cortex, and the lateral prefrontal cortex in adolescents and adults (Cacioppo et al., 2013; Vijayakumar et al., 2017; Wang et al., 2017). The anterior insula and anterior cingulate cortex are thought to serve as a "neural alarm" to select behaviors that motivate social connectedness. Structural and functional alterations of the brain networks involved in social interaction and relational behaviors, as well as in detection of social exclusion or inclusion, help explain some of the individual differences observed between children in their psychosocial functioning and adjustment.

From birth to adulthood, the central and peripheral nervous systems undergo substantial structural and functional changes (Gogtay et al., 2004; Mills et al., 2016; Tau & Peterson, 2009). Neural networks of connected clusters of neurons are formed. Over time and development (based in part on experiences), some connections are strengthened while other connections become weaker (Holtmaat & Svoboda, 2009). Moreover, some brain regions first increase in size early in development before showing gray matter volume reductions (i.e., cell bodies, synapses, and dendrites) and increases in white matter (i.e., myelinated and unmyelinated axons) as those regions become more mature, efficient, and specialized in their functions (Lebel & Beaulieu, 2011; Mills et al., 2016). Developmentally, the most posterior parts of the brain (e.g., occipital lobe) mature earliest and fastest, and most anterior regions (e.g., frontal lobe) are the last to show full maturation (Gogtay et al., 2004). Many of these changes during development are influenced by genes, but those genes function in part in response to environmental inputs (Fox et al., 2010; Peper et al., 2007). Thus, individual differences in environmental inputs interact with genetic differences and very likely shape brain development differently across children (even siblings in the same family). Moreover, although environmental inputs may have effects at any point across the life span, these factors may be most powerful in childhood and adolescence, during the rapid and dynamic brain development that occurs in this period of development (Mills et al., 2016; Mills et al., 2014). In light of this consideration, peer experiences may be particularly important in their influences on brain development prior to adulthood.

Turning to environmental inputs, but before considering peer experiences, it is worthwhile to highlight the largest literature on environmental factors and brain development in childhood—the effects of exposure to chronic stressors (Chen & Baram, 2015; Lupien et al., 2009). The potential effects of stress on brain development during childhood are studied largely in the context of early-life stressors (e.g., premature birth and low birth weight, insecure attachment, maltreatment, malnutrition). Findings from studies of these kinds of stressors show that severe and long-lasting experiences of early-life stress alter emotional and cognitive brain networks involving regions such as the amygdala, hippocampus, and prefrontal cortex (VanTieghem & Tottenham, 2018). For instance, chronic stress leads to hyperactivation of the amygdala but also impairments in PFC functioning, reductions in PFC volumes, and weakening of cortico-limbic connectivity. These stress-related alterations in networks are thought to lead to higher stress sensitivity and increased risk for psychopathology (Bolton et al., 2017).

Experimental studies using animal and in vitro cell models provide insight into the pathway underlying neural alterations following chronic stress as well as evidence for the causality of this stress-exposure mechanism (Arnsten, 2009). Acute social stressors, such as short periods of maternal deprivation or social instability in the home cage, increase the level of stress hormones (e.g., noradrenaline, corticosteroids) in the brain (de Kloet, 2003). An initial rapid release of (nor)adrenaline mobilizes the body and brain to increase vigilance to respond quickly to stressors. This rapid response is followed by a slower increase in corticosteroids, the end product of the hypothalamic-pituitary-adrenocortical (HPA) axis. These corticosteroids subsequently bind to glucocorticoid (GR) and mineralocorticoid (MR) receptors located on the membrane of neurons. Receptor binding results in changes in intercellular processing such as increased energy consumption, cellular metabolism, cell signaling, neuronal connectivity, and neural transmission (de Kloet et al., 2005). The stress hormones and subsequent cellular changes relocate energy away from higher-order cognitive processing toward emotional processing and memory formation. These changes are suggested to be adaptive and allow individuals to respond adequately to such stressors. However, chronic forms of stress have detrimental effects on neural functioning and architecture (Karatsoreos & McEwen, 2013). Chronic stress has been shown to result in dendritic atrophy and lower spine density in the PFC and hippocampus which leads to impaired memory formation and deficits in cognitive functioning. Moreover, the ability of the hippocampus to generate new neurons necessary for learning and memory is impaired, further hampering memory formation (Lucassen et al., 2016). In contrast to the dendritic impairments in the PFC and hippocampus, the amygdala shows increased dendritic spine growth as well as a hyperactivation to stressors (Karatsoreos & McEwen, 2013). These alterations increase stress sensitivity and may be part of the foundation of anxiety, depression, and aggression problems (Fig. 1).

With respect to peer influences, it is now also widely accepted that negative experiences such as rejection, victimization, or bullying serve as important interpersonal stressors in children's lives (Copeland et al., 2013). Experiences of exclusion, peer rejection, or victimization activate the hypothalamic-pituitary-adrenocortical (HPA) axis (Blackhart et al., 2007), a key system implicated in stress responses. Chronic exposure to peer stressors has been associated with dysregulation of the HPA axis and increased risk for internalizing problems such as anxiety and depression (Ouellet-Morin et al., 2011; Peters et al., 2011). Thus, peer exclusion, rejection, or victimization may also lead to neural alterations in accordance with the previously described neural pathway (Fig. 1). The potential effects of peer stressors on brain development and function have mostly been studied with samples of adolescents (Guyer & Jarcho, 2018). Little is known about the middle childhood period.

Positive social experiences with peers may also contribute to brain development. Being accepted by peers or having several high-quality friendships have been shown to be an important predictor for later mental health (Berndt, 2002; Gifford-Smith & Brownell, 2003; Rubin et al., 2006). On a neural level, animal models show that rat pups exposed to enriched social environments (e.g., social housing, high maternal

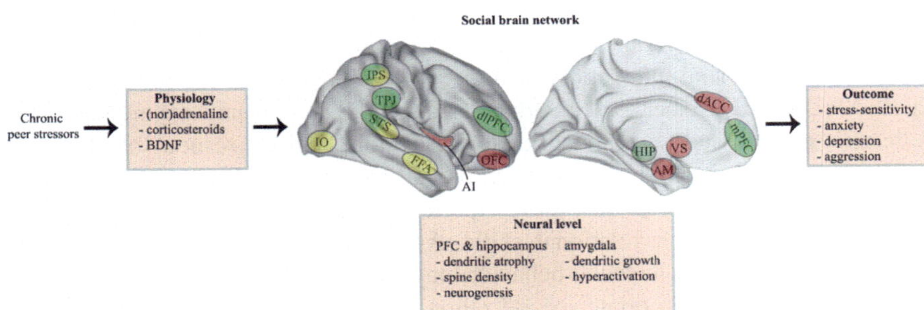

Fig. 1 Theoretical model describing the potential pathway by which chronic peer stressors may lead to negative developmental outcomes via neural alterations in the brain based on results from animal and human studies. Brain regions indicated in yellow indicate regions implicated in the detection node, regions in red are implicated in the affective node, and regions indicated in green are implicated in the cognitive-regulatory node in the social brain network. *BDNF* brain-derived neurotrophic factor, *IO* inferior occipital cortex, *STS* superior temporal sulcus, *TPJ* temporoparietal junction, *IPS* intraparietal sulcus, *FFA* fusiform gyrus, *AI* anterior insula, *OFC* orbitofrontal cortex, *dlFPC* dorsolateral prefrontal cortex, *HIP* hippocampus, *VS* ventral striatum, *AM* amygdala, *dACC* dorsal anterior cingulate cortex, *mPFC* medial prefrontal cortex, *PFC* prefrontal cortex, *TPJ* temporoparietal junction, *STS* superior temporal sulcus

care including sensitivity and secure attachment) show reductions in corticosteroid levels as well as increases in levels of neurotrophins which stimulate neural growth (Curley & Champagne, 2016; van Praag et al., 2000). These growth factors contribute to increased cell survival, growth of new neurons and blood vessels, and increased complexity of the dendritic branches in hippocampal and frontal regions. In humans, the effects of positive experiences on brain development have mostly been studied in the context of positive parenting during childhood. These studies demonstrate associations between positive parenting and total brain volume, hippocampal and amygdala volume, cortical thickness, and brain connectivity (e.g., Kok et al., 2018; Kok et al., 2015; Kopala-Sibley et al., 2018; Wang et al., 2019; Whittle et al., 2014). Behaviorally, exposure to enriched environments or high levels of maternal care may lead to favorable outcomes including lower levels of anxiety and higher levels of social behavior (Baldini et al., 2013; Branchi et al., 2006; Goes et al., 2015; Kok et al., 2018; Lehmann & Herkenham, 2011; Sparling et al., 2018), potentially via neural alterations (Kok et al., 2018). It has yet to be examined how peers during childhood may contribute to children's brain development.

Furthermore, the deleterious consequences of peer stressors—and the potential positive consequences of good peer relationships—may operate not only through alteration of neural networks and functioning but also through social behavior (i.e., experience-dependent synaptic plasticity; Cicchetti, 2002; Feldman & Knudsen, 1998; Hebb, 1949; Kempermann, 2019). If true, having more opportunities to interact with peers—in positive and negative ways—during childhood could have effects on structural and functional neural specialization in brain regions associated with social cognition. Such effects would likely make children more sensitive to social information and help build a more elaborate repertoire of appropriate behaviors to implement during social interactions with peers. For example, children that have many positive social experiences with peers may become better in interpreting mental states or intentions of other children or learn how to guide their own behavior to achieve positive social interactions with peers (Dodge et al., 2003; Lansford et al., 2010; McDonald & Asher, 2018; Parker & Asher, 1987). This improvement may be reflected in increased specialization of these brain regions responsible for this behavior. In contrast, children that are chronically rejected by their peers or who are socially isolated from peers lack these experiences and do not have as many opportunities to practice their relational skills. As a consequence, the relevant brain regions most important to social behavior may lack valuable inputs that contribute to developmentally appropriate specialization. As a result, chronically rejected and neglected children would be at a social disadvantage in part because of alterations in their brain development and functioning—changes that would likely have its influence into adolescence and adulthood (Fox et al., 2010).

Our aim in the current section was to provide a framework that explains how peer experiences may alter children's neurobiology and affect children's development. We suggest that peer experiences may shape regions important for social behavior. Peer stressors experienced by children may alter neural networks implicated in emotion regulation and cognitive control. These changes may be underlying the developmental trajectories of maladjustment so common among youth who are rejected or neglected by their peers. In the next section, we will summarize the neuroimaging literature examining how the brain responds to different peer experiences during childhood, in an effort to deepen our understanding of how peer experiences may shape brain development and social-emotional functioning.

Empirical Evidence

With foundations in theory described, we next turn to the examination of the empirical evidence. In the following section, we review the neuroimaging studies of preschool or middle childhood periods (4–12 years of age). We review neuroimaging studies that assessed behaviors or skills important during daily interactions with peers and then turn to studies that examined neural responses to either peer feedback or social exclusion specifically (two social stressful situations). We end with a description of correlational studies that have examined whether and how individual differences in temperament or peer experiences were associated with brain responses during experimental manipulations involving social situations with peers.

Peer Interaction

Social interactions with peers are highly complex and dynamic. During peer interactions, children need to observe and interpret actions, understand the intentions of their peers, and decide how to respond to their peers. A few neuroimaging studies have looked at how these processes are represented in the brains of children.

Cooperation is a fundamental skill for children that they use during their daily interactions with peers. For example, building a block tower together with a peer requires a child to coordinate behavior with the actions of the other. The child needs to closely observe and cognitively process the actions of others and align those actions with one's own behavior to achieve a common goal. The neural mirroring system may play an important role in processing actions of others and in motivating cooperation (Sartori et al., 2013). In one study, the involvement of the neural mirroring system in cooperation during early childhood was assessed using EEG (Endedijk et al., 2017). First, at age 3 years, cooperation was assessed by observing children while playing a cooperative game together with a same-sex peer. One year later, neural mirroring was assessed using EEG recordings during the observation of videos in which adults were performing different actions such as driving a toy car around. Children that were more cooperative at age 3 showed higher levels of motor system involvement at age 4 during the observation of actions by others, as was indicated by lower power in the beta-frequency band. Lower power in the beta-frequency reflects increased involvement of the motor cortex as well as error monitoring and action updating. These results may thus suggest that for cooperative actions, high levels of neural mirroring and action updating are needed—capacities already in place by 4 years of age.

Another important aspect for successful peer interactions is understanding the emotions, motivations, and behaviors of others—a process referred to as mentalizing. A few recent studies have tried to understand how the brain's mentalizing network responds during dynamic and realistic peer interactions (Alkire et al., 2018; Warnell et al., 2018). Investigators in one study (Warnell et al., 2018) developed an interactive social motivation task in which children aged 8–12 years were presented with sentences about themselves such as "I play soccer" or "I like French fries" during an fMRI session. Participants were asked to indicate whether these sentences were true or not and then had to wait for a reply. Replies could be either from a peer (peer trial) or from the computer (computer trial). Children were told that this peer was an age- and gender-matched peer, but the replies were simulated. To capture the interactive nature of peer interactions, another manipulation was added such that replies from peers could either be engaged ("Me too" or "I didn't pick that") or nonengaged because they were performing another task ("I'm busy"). Computer replies could also be engaged ("Match" or "Mismatch") or nonengaged ("Disconnected"). Both initiating an interaction and receiving a reply from peers (compared to a computer interaction) resulted in increased activation in the ventral striatum, a region implicated in reward processing. This suggests that interacting with peers is considered rewarding for children. Interestingly, activation in the ventral striatum was not stronger for older children compared to younger children; peer interactions may be equally rewarding across middle childhood. In contrast, receiving an engaged reply from a peer was related to increased activation in the dorsomedial prefrontal cortex—and this response was stronger for older than younger children. The dorsomedial prefrontal cortex is part of the affective node and has been implicated in processing information about self and others (Pfeifer et al., 2007). Other mentalizing areas (i.e., temporoparietal junction, superior temporal sulcus) also showed increased activation when interacting with a peer, relative to a computer—and again, the response was stronger among older children. Thus, socially relevant information about peers is processed by the socio-cognitive and reward network, with some of aspects of the neural activity and socio-cognitive skills continuing to develop across middle childhood.

A different interactive mentalizing fMRI task was used in another study of 8–12-year-olds (Alkire et al., 2018). Children received a hint about a chat partner or a fictional computer character and had to predict what this peer or character would pick as a choice (mental condition) or guess which option would match a certain situation (nonmental condition). For example, "Sue likes to read" would require an answer indicating that she will pick a long novel (mental), while a hint saying "Sue has big feet" would require an answer related to needing bigger shoes (nonmental). After this guess, participants received feedback whether their responses corresponded with those of the peer or character. Results showed that guessing the response of a peer resulted in higher activation in the mentalizing network, compared to guessing the responses to a fictional character. Similar to the study by Warnell et al. (2018), the dorsomedial prefrontal cortex and temporoparietal junction were more engaged in the social interactions with peers and not during mentalizing about a computer character. Moreover, social interactions with peers were considered more rewarding as shown by increased activation in the reward network (e.g., striatum, medial orbitofrontal cortex) compared to guessing reactions to a fictional computer character.

Regarding development, Warnell et al. (2018) found increased activation in the socio-cognitive network with age, but Alkire et al. (2018) found *decreased* activation for older children during mentalizing about peers relative to fictional characters. However, this difference in age effects between studies may be driven by differences in the nonsocial condition. Whereas the nonsocial condition in Warnell et al. (2018) was a computer, the nonsocial condition in Alkire et al. (2018) was a fictional character. The lower activation for older children in the study by Alkire et al. (2018) may be driven by developmental improvements in mentalizing activity for the fictional character compared to the peer, whereas mentalizing about a computer may not improve with age. Despite these distinct age effects, both studies showed that social interactions with peers are considered rewarding during childhood and are processed in socio-cognitive regions important for mentalizing.

The rewarding nature of social interactions with peers may influence how children behave. Prior studies in adolescents showed that the mere presence of a peer results in increased risky behavior and enhanced activation in the reward network (striatum) of the brain (Albert et al., 2013; Chein et al., 2011). Yet, this phenomenon has received relatively little attention during childhood. One study of 10–14 year olds ($M = 12$ yrs.; Hoffmann et al., 2017) demonstrated that peer presence may also influence how the brain reacts to risk-taking situations. Children completed a computerized risk-taking task (Balloon Analogue Risk Task; Lejuez et al., 2002) in the fMRI scanner while being observed by a peer. Youth took fewer risks and showed increased activation in the amygdala and medial temporal lobe during peer observation compared to completing the task alone. The amygdala has been implicated in processing affective salient information and loss aversion (Janak & Tye, 2015). Thus, when children are observed by peers, risks may become more salient and aversive which may reduce risk-taking behavior.

A different condition in Hoffmann et al. (2017) involved peers encouraging the child to take more risk in the next trial. Peer encouragement resulted in higher levels of risk and greater activation in the inferior frontal gyrus (IFG), compared to trials in which the participants was only being observed by peers. The increased IFG activation may relate to integrating the information from peers in deciding and executing a behavioral response (Dippel & Beste, 2015). Results from Hoffmann et al. are distinct from studies of adolescents. That is, while adolescents took more risks during the mere presence of peers (Chein et al., 2011; Gardner & Steinberg, 2005), children actually became more risk-aversive when being observed by peers and only took more risks when peers verbally motivated them to take more risks. Furthermore, while the mere presence of peers elicited activation in the brain's reward regions (i.e., ventral striatum) (Chein et al., 2011), this childhood sample showed increased activation in a region implicated in salience processing. Thus, the effects of peers appear to be different across childhood and adolescence both on a behavioral and neural level, thereby underscoring the importance of studying peer processes across different developmental periods.

In sum, neural evidence for children's motivation to interact with peers is shown by increased activation in brain reward regions during peer interactions. In addition, the brain imaging evidence shows neural activation patterns that correspond with children actively monitoring the actions of others, trying to infer mental states of their peers, and adapting their behavior based on input from their peers. The literature long ago established the salience of peer interactions based on behavioral evidence; the more recent neuroimaging findings deepen that evidence by including neural indicators.

Peer Feedback

One of the most salient and important peer experiences—in middle childhood as well as adolescence and adulthood—is being socially evaluated by peers and receiving feedback on attributes, behaviors, or beliefs (Somerville, 2013; Westenberg et al., 2004). This feedback can be rewarding or aversive. Receiving negative feedback from peers is one of the most frequent peer stressors in childhood and adolescence. Examples include a peer saying that he or she does not like the child's clothes or a teenager receiving no "likes" on their social media post. These events can be highly salient for children and adolescents, as part of the social information being gathered to determine their social inclusion within a peer group.

Neural processing to peer feedback has been examined in adolescence (Gunther Moor et al., 2010; Guyer et al., 2011; Guyer et al., 2009; Somerville, 2013) and young adulthood (Davey et al., 2010; Somerville et al., 2006), but investigation of these processes in middle childhood has only begun recently. Typically, neural responses to peer feedback are elicited using a social feedback task. Days or weeks prior to the fMRI session, participants are asked to fill out a form with personal information about their favorite movie or sport, and other likes and dislikes. The participants are then led to believe that their profiles will be reviewed by other peers. During the fMRI session a few days or weeks later, participants are shown pictures of age-matched peers and are also presented with feedback about how that peer felt about the participant's profile. This peer feedback can be positive, negative, or neutral. Among adolescents, negative feedback is associated with increased activation in the salience network including the amygdala, anterior insula, and medial prefrontal cortex (Somerville, 2013). Positive feedback is rewarding, indicated by increased activation in the striatum.

Social feedback tasks are now being used in brain imaging research in middle childhood. In one study, 7–10-year-olds completed a social network aggression task (Achterberg et al., 2018; Achterberg et al., 2017). Children filled out profiles at home prior to the fMRI session. During the fMRI session, children received positive, negative, or neutral feedback from peers. However, in contrast to the social evaluation task described previously, this task allowed children to respond to the peer feedback with a loud noise blast that would be presented to the peer. The length of the noise blast was used as an indication of the level of aggression toward that peer.

Results of this study (Achterberg et al., 2018; Achterberg et al., 2017) showed that noise blasts were longer after negative feedback, followed by neutral feedback, and the shortest noise blasts were given after positive feedback. With regard to neural activity, peer feedback—regardless of valence—resulted in activation in the anterior cingulate cortex. In contrast, only negative feedback was associated with increased activation in the medial prefrontal cortex, and only positive feedback was associated with increased activation in the caudate. One interpretation is that the anterior cingulate cortex is involved in processing all socially salient information, the medial prefrontal cortex is more specific in its processing of socially threatening information, and the caudate region is more specific in its processing of socially rewarding information. Interestingly, the neural patterns in the studies in middle childhood are similar to those reported in adolescence (Davey et al., 2010; Gunther Moor et al., 2010; Guyer et al., 2011; Guyer et al., 2009). Thus, the neural components of processing salient social information from peers already are in place by middle childhood.

But is it possible that those processes are evident even earlier in development? One study examined this question in a sample of 4–6-year-olds, using an adapted, age-appropriate version of the social aggression network task (van Wijk et al., 2019). Two weeks before the EEG lab visit, children were asked to pick one out of five cuddly animal toys as their favorite and received this toy at home to become highly familiar with it. During the lab visit, EEG was recorded while children received feedback from hypothetical peers on their cuddly animal (e.g., "Your cuddly animal is stupid"). Instead of delivering a noise blast to the peer, children were told that the peer had ten balloons and the target child could destroy balloons by pressing a button. Longer button presses resulted in more destroyed balloons and were used as a measure of individual differences in aggression. Neural processing of feedback was assessed by examining EEG alpha power frontal lobe asymmetry, an indicator of approach and avoidance motivation and behavior (Kelley et al., 2017). Negative feedback resulted in longer button pressing compared to neutral and positive feedback, suggesting that these young children were aware of and affected by negative peer evaluation. However, there were no associations (direct or indirect, i.e., mediating) between asymmetry and feedback condition or button pressing. Thus, evidence of behavioral responses to peer negative evaluation were evident among 4–6-year olds, but the null finding using EEG leaves open questions about whether and how neural processing is involved at this young age.

In sum, studies of adults and adolescents have established neural response patterns to negative and positive peer evaluative feedback. Although results from studies of children are preliminary, similar patterns are evident in middle childhood. It remains to be seen whether those patterns are established even earlier in development (e.g., during the toddler or preschooler years).

Peer Exclusion

Another peer stressor that some children may experience during their interactions with peers in school is exclusion from social activities. Examples include the following: when a child does not receive an invitation to a party of a classmate, but all the others are invited, and when a group of children is playing a game at the schoolyard and one child is not asked to join and is even ignored when asking to be part of the game. These experiences can be very stressful and emotionally painful for children and may be evident in changes at the neural level as well. However, there is sparse research in childhood on neural processing of peer social exclusion—most of the available research has included adolescent and adult participants.

One common paradigm employed by peer exclusion studies that measure neural responses to social exclusion is the Cyberball, an "online" ball-tossing game (Williams & Jarvis, 2006). Participants are led to believe that they are playing this ballgame with two other players. In reality, the game is preprogrammed by the experimenter. During the Cyberball game, participants first participate in a fair play round in which all players receive the ball as often as the others. However, as the play proceeds, the participant becomes excluded from the game by the two other players and no longer receives the ball. Neuroimaging studies in adolescence and adulthood have shown that being excluded from this ballgame is experienced as distressing, as shown by temporary decreases in mood and need satisfaction levels and increased activation in the medial prefrontal cortex, insula, anterior cingulate cortex, precuneus, and lateral prefrontal cortex (Cacioppo et al., 2013; Vijayakumar et al., 2017; Wang et al., 2017). Whereas the activity in the insula, medial prefrontal cortex, and anterior cingulate have been related to the negative affect induced by social exclusion, the activity in the lateral prefrontal cortex has been implicated in regulation of this negative affect (Eisenberger et al., 2003; Masten et al., 2009). Interestingly, studies have adapted parts of this game by using shorter durations and alternating social inclusion and exclusion rounds, and yet, the neural processes observed during social exclusion have been very similar across the game variants (Vijayakumar et al., 2017). Thus, neural processing of social exclusion in Cyberball appears to be very robust.

The Cyberball task has also been used in a few EEG studies to examine neural sensitivity to social exclusion in preadolescent children (Crowley et al., 2010; van Noordt et al., 2015). Results showed that early and later stages of social exclusion elicited increased neural activity among children aged 8–12 years old. During early periods (264–656 ms), an enhanced positive ERP component, resembling the P300 component, was found at the posterior site (Crowley et al., 2010) and higher theta oscillations in medial frontal sites (van Noordt et al., 2015). These findings were interpreted to reflect signaling conflict and salience of the exclusion experience. During later periods of social exclusion, Crowley et al. and van Noordt et al. also both reported larger ERP negativity and increased theta oscillations in medial frontal regions. Also, in both studies, the later ERP components were associated with the amount of distress children reported. Together, these findings show that by age

8–12 years, children show increased neural sensitivity during social exclusion that reflects detection of exclusion and neural modulation of emotional responses.

Differences in neural processing of social exclusion by strangers versus friends have also been studied. In an EEG study by Baddam et al. (2016), children aged 8–12 years and their best friends were invited for participation. Both the child and best friend played the Cyberball task in a separate room. Participants were told that they would play the Cyberball task together with their best friend and a stranger, but in reality the task was preprogrammed by the experimenter. Children first played a social inclusion round in which the ball was tossed fairly toward all players (child, best friend, and stranger). In the second round, the child was for the most part excluded by the stranger and best friend. Tosses from the friend toward the stranger during the exclusion block were considered as an exclusion trial by a friend, and tosses from the stranger to the friend were considered an exclusion trial by a stranger. Results showed that exclusion by a stranger was associated with larger P2 responses and higher slow wave activity in medial frontal sites compared to exclusion by a friend. The larger P2 responses may indicate higher attentional allocation to rejection by strangers compared to friends. Increased slow wave activity may indicate processing of aversive stimuli, arising from the aversive nature of exclusion by strangers compared to friends.

One potential explanation for the increased sensitivity to social exclusion from strangers may relate to the social competition between peers and the instability of friendships during middle childhood (Hartup, 1996; Schneider et al., 2005). In this Cyberball task, strangers threw the ball toward participant's best friends and excluded the participant. As such, strangers may intrude in the friendship, which is a highly salient experience that may lead to distressed feelings and increased neural sensitivity to exclusion among the participants.

Social exclusion experiences also include passively observing other individuals being excluded, which also may be very distressing for children (Saylor et al., 2013). A child can decide to compensate for the exclusion by including the excluded child. Compensation for peer exclusion may be socially challenging and may influence the participant's own inclusionary status. To examine this prosocial compensation behavior, the Prosocial Cyberball Game (PCG) was developed and tested (Riem et al., 2013). In the PCG, the participant is playing an online ball-tossing game with three other virtual players. During the first round, all players receive the ball an equal amount of time (i.e., fair play). During the second round, during the unfair round, one player is excluded by the two other players—but in contrast to the original Cyberball, the participant still receives the ball from the other players. The participant can decide whether to include the excluded player by tossing the ball toward the excluded one (the so-called prosocial compensation). Prosocial compensating behavior is assessed by comparing trials in which the participant tosses the ball toward the excluded player, with trials in which the participant is tossing the ball toward the excluders. Higher percentages of tosses toward the excluded player during the unfair round compared to the fair round are considered compensating behavior. Two recent studies examined the behavioral and neural responses of this compensating behavior among 7–11-year-olds (van der Meulen et al., 2017; van der

Meulen et al., 2018). Results of both studies showed that children compensate for the social exclusion by tossing the ball toward the excluded player during the exclusion round more often than during the fair round. The neural correlates were less clear, however. In the 2017 study, no neural correlates were found, but in the subsequent 2018 study, compensating behavior was associated with increased activation in the posterior cingulate cortex/precuneus and with lower insula activity. Activation changes in these brain regions suggest that children who show stronger compensating behavior are striving to infer the mental states of others and are modulating affective responses while observing the peer exclusion.

When observing a peer being excluded, a child may also worry or be concerned about becoming excluded themselves. This concern about becoming excluded also was examined in van der Meulen et al. (2017), and (2018), by comparing trials in which the other players excluded the participant with trials in which the participant *did* receive the ball from the other players (referred to as *self-exclusion*). Self-exclusion resulted in increased activation in the inferior frontal gyrus, insula, hippocampus, caudate, amygdala, and occipital gyrus. These neural correlates overlap partly with previously described neural correlates of social exclusion in adolescents and adults and are linked to affective responses (Cacioppo et al., 2013; Vijayakumar et al., 2017; Wang et al., 2017). Also, being included by the excluders relative to self-exclusion resulted in increased activation in the supplementary motor area, precentral gyrus, middle frontal gyrus, anterior cingulate cortex, supramarginal gyrus, middle cingulate gyrus, and caudate. Again, results only partly overlap with findings in adult studies such as the anterior cingulate cortex and caudate (Dalgleish et al., 2017; Perini et al., 2018). These regions are implicated in attentional processing, appraisal of events, and reward processing thereby indicating the importance of being included in childhood (Delgado, 2007; Shenhav et al., 2013). Heritability analysis showed that the neural responses to self-exclusion and prosocial compensating behavior were related to nonshared environment and measurement error. This was surprising, given that prosocial behavior has been shown to be heritable in childhood (Gregory et al., 2009; Knafo & Plomin, 2006). However, prosocial behavior in prior behavior genetic studies has usually been measured with questionnaires; in contrast, the prosocial compensating behavior being examined in the studies by van der Meulen et al. (2017, 2018) may reflect state-like responses to social exclusion. In other words, variation between youth in prosocial compensating behavior may be less genetically influenced than overall prosocial tendencies. In addition, the blood-oxygen-level-dependent (BOLD) response underlying the fMRI signal is quantitatively “noisy”. Heritability studies using BOLD have not yet been able to separate out specific measured environmental influences from measurement error. More research is necessary to interpret the nonshared environment finding.

Together, results from the studies described in the current section demonstrate that children between 7 and 12 years of age show neural sensitivity to social exclusion. These neural correlates overlap to some degree with the neural correlates found in adolescent studies (e.g., inferior frontal gyrus, insula). Furthermore, social exclusion by strangers seems to be more distressing and salient (in terms of neural sensitivity and processing) than social exclusion by friends. In addition, some children who observed someone else being excluded compensated for the exclusion by behaving prosocially toward the target; this variation was associated with increased activation in social-affective and mentalizing brain regions. Although the neural correlates of compensating behavior in childhood were inconsistent across studies, this new line of research lays a foundation for future imaging studies that will allow us to deepen our understanding of how social exclusion experiences are processed and responded to, by children.

Individual Differences in Temperament or Prior Peer Experiences

The studies described so far have shown that children are sensitive to peer stressors, and these peer stressors elicit brain activation in affective and cognitive control regions. However, those studies have not considered individual differences in children's and adolescents' temperament and peer experiences prior to the studies' assessments, yet these may matter a great deal. Not all children are sensitive to peer social experiences to the same extent. According to the differential-susceptibility hypothesis (Belsky & Pluess, 2009) and the diathesis-stress model (Boyce & Ellis, 2005), some children show higher sensitivity to stressors (such as negative peer experiences), whereas other children may be relatively unaffected by the presence of (chronic) stressors. The differential-susceptibility hypothesis further states that children who are more sensitive to negative social contexts may flourish in very supportive and positive contexts (Belsky & Pluess, 2009). Thus, there are likely to be subgroups of children who benefit in particular from both the absence of stressors (e.g., peer rejection) and the presence of positive enriching factors (e.g., having strong friendships). It is important to consider these individual differences in sensitivity to peer environmental factors to understand possible resilience factors that can be targeted for interventions or preventions. Therefore, we next consider recent fMRI studies that have examined how individual differences in prior temperament or peer experiences during childhood are associated with sensitivity to peer interactions.

In one neuroimaging study of 11-year-olds (Jarcho et al. (2016), children who were either high or low in social reticence (i.e., silent, withdrawn) participated in a fMRI session while performing a peer feedback task called the virtual school paradigm (Jarcho et al., 2013). The virtual school paradigm was developed as a feedback task that would be similar to situations in classrooms and capture the dynamic nature of social feedback processing involving anticipating and receiving peer feedback from others who have reputations as being nice, mean, or unpredictable. Anticipating feedback from peers with a mean or unpredictable social reputation may be even more distressing than anticipating peer feedback from anonymous peers whose reputations are unknown (as is the case in the Cyberball studies described earlier). The virtual school paradigm consists of two sessions. Days prior to the experiment, participants are told that they will participate in a game in which they are placed in a virtual school environment. To increase the salience of this experience, children are asked to create their own avatar and also fill out a personal profile with their interests. During the second visit, prior to the fMRI assessment, children are introduced to several virtual classmates and receive information on their social status—two who are nice, two who are mean, and two who are unpredictable in their behavior. Children complete the virtual school paradigm inside the scanner. Every trial starts with the image of a virtual classroom with the different peers (nice, mean, or unpredictable). One peer starts typing a feedback response and participants see the text ("Is typing…") in a text balloon close to this virtual peer (the "anticipation" phase). Children then see the feedback of the peer which is either from a set of pre-generated responses (e.g., "you're lame") or containing information specific to the child's previously completed personal profile. Following feedback, children can respond to the feedback from predetermined responses that are either nice, mean, or avoidant (no response).

Results showed that children who were high in social reticence showed increased activation in the dorsal anterior cingulate cortex and mid-to-anterior insula during the anticipation of feedback from an unpredictable peer (Jarcho et al., 2016). Given the role of these regions in salience and affective processing, these findings suggest that children high in social reticence find anticipating unpredictable feedback more salient and distressing than children lower in social reticence. Moreover, a weaker functional coupling was found between the insula and regions involved in inhibitory control (ventromedial prefrontal cortex, premotor cortex) for children high in social reticence. During negative feedback from unpredictable peers compared to mean peers, amygdala activation decreased for reticent children but increased for non-reticent children. The amygdala may serve a role in valence signaling, which may interact with the level of predictability of peer feedback (McHugh et al., 2014). Furthermore, non-reticent children may process unpredictable negative feedback as more salient as it was not expected, while children high in social reticence may exhibit abnormal processing during these situations.

The virtual school paradigm was also used in another study examining how childhood wariness (i.e., fearful or passive behavior in social situations) and victimization in school relate to neural activation during peer feedback, in a sample of 11-year-olds (Jarcho et al., 2019). No behavioral differences in responses to feedback were observed for children that were victimized or scored high on wariness. However, highly victimized children that scored high on childhood wariness showed greater activation in the amygdala, striatum, and left insula during positive feedback from unpredictable peers. In addition, the degree of amygdala activation in highly victimized, highly wary children was positively associated with self-reported levels of social anxiety. These effects were found for *positive* feedback, and the study by Jarcho et al. (2016) showed *dampened* amygdala activation for unpredictable *negative* feedback among reticent children. The amygdala may play a role in valence signaling, novelty processing, and reinforcement learning in unpredictable social situations (Janak & Tye, 2015). Children high on social reticence may frequently experience negative feedback from unpredictable peers, and positive feedback from others may be less common for youth high on victimization and wariness. Findings from these studies show the importance of considering both personal risk factors (i.e., wariness) and social contextual risk factors (i.e., victimization) to understand differential sensitivity to peer interactions and potential links with maladjustment (i.e., social anxiety).

Besides showing correlations with differential brain activation during peer feedback situations, behavioral inhibition and social reticence may also relate to the child's behavior during socially stressful situations. In a study by Lahat et al. (2014), 7-year-olds performed a cognitive control task (i.e., Flanker) while EEG was recorded to assess cognitive conflict using the N2 component. Following the EEG recordings, participants played a real-life ball-tossing game with an unfamiliar experimenter during which the child was excluded after a few minutes. Children's behavior in response to the social exclusion was observed and coded. Children that were behaviorally inhibited and had higher N2 amplitudes during the cognitive control task showed higher socially withdrawn behavior and lower assertiveness in response to a social exclusion experience. Behaviorally inhibited children with lower N2 components did *not* show this withdrawal behavior. In general, higher levels of executive function have been linked to better peer relationships from early childhood through adolescence (Holmes et al., 2016). However, results from Lahat et al. show again that personal risk factors (e.g., behavioral inhibition, social reticence) may influence social behavior in response to peer feedback. Children who are highly inhibited may not be able to adaptively and flexibly respond to socially stressful situations when they also show high levels of cognitive control over their behavior. This may contribute to maintaining socially withdrawn behavior that influences subsequent social-emotional development.

Prior experiences with peers may also sensitize children to expect new rejection events—sometimes called *rejection sensitivity* (London et al., 2007). Behavioral evidence has shown that children high in rejection sensitivity demonstrate an increased vigilance for hostile rejection cues and respond more aggressively to new rejection experiences (Dodge et al., 2003; Lansford et al., 2010). Adolescents with a history of peer rejection or victimization also show this sensitivity, including distinct neural changes that reflect greater rejection sensitivity to new rejection events (Rudolph et al., 2016; Will et al., 2016). Specifically, youth with a history of peer rejection or victimization show increased activation in the dorsal anterior cingulate cortex during new rejection events, compared to those without such history. However, as mentioned in the previous section, neural correlates of social exclusion experiences only partly overlap when comparing childhood and adolescent samples (Cacioppo et al., 2013; van der Meulen et al., 2017, 2018; Vijayakumar et al., 2017; Wang et al., 2017)—thus, the neural patterns associated with exclusion experiences for those with histories of peer problems may be distinct for children and teenagers.

There have been only two studies we know of that have examined whether and how long-term prior experiences of peer rejection associate with neural rejection sensitivity in childhood. In an interpersonal feedback brain imaging study with 11-year-olds who had, or had not, experienced chronic peer rejection (Lee et al., 2014), children were asked to provide answers to unsolvable puzzles inside the MRI scanner. They then received interpersonal feedback in the form of photos of facial expressions of others (i.e., positive, negative, neutral). Peer-rejected children showed increased activation in the "social pain" network (Eisenberger, 2012) including the orbitofrontal cortex, ventrolateral prefrontal cortex, hippocampus, and amygdala during negative feedback. During positive and neutral facial feedback, increased activation in the inferior occipital gyrus was found for these rejected children. Thus, peer rejection in school was associated with heightened sensitivity to negative feedback after trying to solve an unsolvable puzzle; this was *not* the case for 11-year-olds without a peer rejection history. As found in other studies described earlier, increased amygdala activation was found during negative peer feedback, showing the importance of this region in processing stressful peer experiences during childhood. It is possible that children who experience peer rejection in school may be more socially vigilant and anxious, which may underlie the observed increase in amygdala activation. Again, results from this study align with other studies on social reticence, behavioral inhibition, and victimization showing how prior experiences with peers pertain to individual differences in neural sensitivity to peer interactions during childhood.

In another study, Asscheman et al. (2019) examined how prior experiences of peer rejection were associated with neural sensitivity to a new peer stressor, among 8–12-year-old boys. Children's history of peer rejection was determined based on peer nomination measures assessed in elementary school 3 years before the fMRI study. Boys with stable levels of low or high peer rejection and peer acceptance over these 3 years participated in a fMRI study in which participants were excluded during the Cyberball. Peer-rejected boys showed increased activation in bilateral dorsolateral prefrontal cortex as well as supramarginal gyrus compared to peer-accepted boys. These results thus show that chronic peer rejection during middle childhood is associated with increased activity in regions associated with emotion regulation, attentional control, and social cognition. Like Lee et al. (2014), there was no associated change in dorsal anterior cingulate cortex activation—a brain region that has been linked with peer exclusion experiences in prior imaging studies of rejected or victimized adolescents (Rudolph et al., 2016; Will et al., 2016).

In sum, the relevant childhood brain imaging literature shows that child characteristics as well as prior real-life peer rejection experiences are associated with differential sensitivity to stressful peer feedback and exclusion behavior. Although these studies are cross-sectional, if future research demonstrates stronger causal evidence, it would mean that prior peer experiences and behavioral risk factors may shape how the brain responds to new peer experiences and influence the pathway between peer stressors and psychopathology (e.g., social anxiety). Although some similarities in neural correlates were found with the imaging studies of adolescents, some neural correlates were not found in childhood samples (i.e., dorsal anterior cingulate cortex). Thus, peer stressors may be processed differently in childhood, compared to during and after puberty. More research is necessary in middle childhood and especially in early childhood when peer relationships and social skills are first emerging—a developmental period when prevention and intervention may be most effective (Fox et al., 2010; Shonkoff & Levitt, 2010; Wachs et al., 2014).

Future Directions

The studies presented in the current chapter demonstrate that progress is being made in our understanding of peer influences on brain function and development during childhood. However, many outstanding questions remain. In closing, we provide some directions for future research to advance the field.

First, longitudinal studies are needed, but there are challenges to be overcome for that work to proceed. Environmental influences such as peer relationship histories may alter how the brain responds to new social information, but the empirical studies to date are cross-sectional; as a result, we were not able to interpret likely neural changes over time and development. It remains to be seen how individual differences in brain responses to social information emerge and change with development and how these individual differences may explain longer-term developmental outcomes such as behavioral and emotional problems in later childhood, adolescence, and early adulthood. Longitudinal fMRI studies will give invaluable insight into the variability of the BOLD signal on different time points across development. This could address questions related to trait- and state-like brain activation patterns that might permit stronger inferences from the existing cross-sectional literature (i.e., ruling out cohort effects). Moreover, longitudinal studies allow mapping of the dynamic features of developmental trajectories (e.g., timing of nonlinear changes; rate of changes) of brain function. It may be individual differences in those dynamics (i.e., delays, velocity), rather than more general variance in functional and structural brain differences that are most predictive of outcomes (Shaw et al., 2010). For example, children with ADHD show delayed and slower cortical maturation across childhood, which may result in the cognitive control deficits found in these children (Shaw et al., 2007). Although longitudinal brain imaging studies with children are challenging, there remains a need for such studies to elucidate likely causal effects of peer experiences on brain development and developmental outcomes.

Another issue to address in future research concerns whether the tasks (and resulting scores) used in imaging studies meet the standard of invariant measurement required for comparing means and variances across multiple time points in development (Telzer et al., 2018). For example, brain activation in specific regions of the brain during a cognitive control task may show an average increase with age across development in childhood. However, this increase could indicate an improvement in cognitive functioning or simply reflect an improvement in neural signal-to-noise ratio in BOLD signal at later ages when behavioral performance is better (i.e., less noise and therefore improved signal estimation). One potential solution is to adapt tasks over development, so that the overall level of difficulty (i.e., error rates) for each age point remains constant over time. An even more fundamental measurement issue in longitudinal brain imaging studies is the lack of evidence for strong test-retest reliability of BOLD scores from fMRI (Herting et al., 2018). The solution likely will require establishing estimates of short-term test-retest reliability over the course of weeks and then adjusting longitudinal stability estimates for reliability of measurement. This statistical approach has been shown to be useful in other fields

(Heise, 1969). Applying this approach to fMRI will be time-consuming and expensive, yet the field would benefit from more extensive collaborations to assess neural measures with large samples of children across shorter intervals. Finally, there are measurement issues for social behavioral tasks such as Cyberball. Debriefing is essential because of the deception that is involved, making repeated assessments intrinsically different from the initial deception assessment. Future research will need to address this and other measurement challenges when conducting longitudinal studies.

Second, our knowledge on peer socialization and brain development requires studying not only brain function but underlying anatomical structure and connectivity (Wang & Olson, 2018). Changes in brain function may already occur after a single experience. Repeated and enduring co-activation of neurons results in structural connectivity changes ("neurons that fire together, wire together"; Hebb, 1949). Therefore, changes in gray and white matter structures and connectivity during development will reveal more useful information about the long-lasting changes arising from prior experiences. A recent cross-sectional study showed that victimized adolescent boys exhibited lower ventrolateral prefrontal cortex volumes (du Plessis et al., 2019). In addition, social interactions are highly complex and dynamic and require fast real-time processing and integration of information that depends heavily on white matter structures (Kennedy & Adolphs, 2012). Studying white matter connectivity may reveal some of the underlying deficits in social competence that may help explain peer difficulties. For example, social anxiety symptoms may be better explained by individual differences in white matter structures compared to gray matter structures (Whitfield-Gabrieli et al., 2015). Furthermore, it has been theorized that stress during early life accelerates structural brain changes in emotion networks (i.e., stress acceleration hypothesis; Callaghan & Tottenham, 2016). Faster structural brain changes may be detrimental for children and adolescents, as this acceleration may affect how much time these children have to learn and refine behavioral skills necessary for the psychosocial challenges found during adolescence (Ge & Natsuaki, 2009). Rapid brain maturation may thus increase children's vulnerability for psychopathology. Thus, future studies should assess structural brain maturation in relation to peer environmental factors and adjustment outcomes.

Third, future research will benefit from inclusion of positive peer experiences as potential "buffers" of peer stressors in development. For example, one brain imaging study of adolescents (Telzer et al. (2015) showed that adolescents who experience high levels of peer conflict during their daily lives also showed higher levels of risk-taking behavior and increased brain activity in regions associated with affective processing (e.g., insula, ventral striatum). However, having supportive peer relationships had a buffering effect. Similarly, in a study of adults, neural responses to a social exclusion experience were dampened when participants were reminded of their attachment figure (Karremans et al., 2011).

Fourth, social relationships in childhood go well beyond peer relations at school. Social interactions with peers are not static and do not occur in a vacuum; children's social lives are dynamic and intersect with social lives involving siblings and adults. Children continue to have attachment relationships with parents beyond early childhood, and nonparental adults (e.g., teachers) play an important role in children's social and emotional development. In addition, children spend time with their peers and siblings outside of school settings, in their homes, neighborhoods, and other settings. According to the stress-buffering hypothesis, social support from multiple relationships with other children and with adults may mitigate the impact that stressors have on health and functioning (Cohen et al., 2000). Future research would benefit from considering how these other social relationships interact with negative and positive peer relationships in school.

Fifth, future studies should take advantage of progress in the use of functional near infrared spectroscopy (fNIRS) (Wilcox & Biondi, 2015). This imaging technique has several advantages over other neuroimaging techniques such as EEG and fMRI (Ferreri et al., 2014). Movement artifacts can better be dealt with in fNIRS than with fMRI which is a great advantage when measuring brain activity in physically active and "fidgety" children (Lloyd-Fox et al., 2010). Relatedly, most EEG studies use event-related potentials (ERP) to understand the neural processing of peer stressors in young children. However, ERP signals require many trials with high-quality signals which may be very challenging to collect in these young children due to high motion. Another advantage of fNIRS is that, like ambulatory EEG hardware, fNIRS hardware can be used in ecologically valid settings (e.g., schools) so the data that are collected can be more natural for children compared to the confined setting of MRI scanners. However, there are limitations with fNIRS. fNIRS has better spatial resolution but poorer temporal resolution than EEG. Also, fNIRS has better temporal resolution than fMRI, but unlike fMRI, it can only measure about 1 cm into the surface of the cortex. Peer environmental experiences may be associated with emotional and reward systems deep in the brain, and fNIRS cannot measure those responses. With advantages and disadvantages compared to EEG and fMRI, fNIRS is highly suitable for use with infants, children, and adolescents in cross-sectional and longitudinal designs.

Sixth and finally, an interesting and novel approach is to investigate a potential role of the gut microbiota in the link between peer experiences and brain development. Gut microbiota may modulate brain function and development via the microbiota-gut-brain axis (Brett & de Weerth, 2019), and bidirectional links exist between intestinal microbiota composition and brain function (Collins et al., 2012). Dysbiosis in the gut microbiome has been linked to alterations in brain development (Rogers et al., 2016). Moreover, pre- and postnatal stress may alter the composition of the gut microbiome (O'Mahony et al., 2017; Zijlmans et al., 2015). It has yet to be investigated but peer stressors during development may also potentially influence the gut microbiome.

Implications and Conclusions

The research presented in the current chapter will one day inform prevention and intervention practices as well as public policy. Increasing our understanding of how various risk factors differently influence brain development and psychosocial adjustment in childhood (as opposed to adolescence or adulthood) may allow earlier identification of those individuals who are most sensitive to peer stressors and are at most risk to develop behavioral and emotional problems. Relatedly, knowledge about the buffering effects of positive features of other social relationships (e.g., parents, teachers, friends) on neural sensitivity to peer stressors could be used to prevent some of the negative outcomes for children as well as adolescents who are experiencing peer difficulties.

Moreover, new avenues of research arising from neurocognitive studies may provide new opportunities for affordable and simple interventions. For example, although the research is new and requires much more study, there is mounting evidence of bidirectional links between brain development and gut microbiota. This may be an effective target for interventions for promoting healthy brain development in the presence of peer stressors, such as changing the composition of the gut microbiota with probiotics (Brett & de Weerth, 2019; Wang et al., 2018). Lastly, longitudinal studies provide more conclusive evidence than cross-sectional studies of links between peer stressors and brain development. That growing literature suggests that prior peer experiences can become "embedded" in children's (neuro)biology and may influence subsequent brain development as well as behavioral, cognitive, and emotional functioning. The transactional process between peer problems and brain development may lead to significant neural alterations that set the stage for psychopathology. Schools will do well to include social and emotional learning as part of their curricula (in addition to standard academic subjects) from kindergarten onward, to promote healthy development of the brain and body during childhood and beyond (Greenberg et al., 2017).

In conclusion, our goal in the current chapter was to outline the progress being made in the field of social developmental neuroscience by reviewing the literature on neural responses to peer experiences (i.e., interactions, feedback, exclusion) during the elementary school period (4–12 years old) as a comparison to the more extensive literature on adolescents. The theoretical framework is that peer experiences may lead to specific and long-lasting changes in neural structures and functioning across development—changes that may enhance risk for psychosocial and health problems later. The literature suggests that children show neural processing during peer interactions and in response to peer stressors in affective, socio-cognitive, and control networks—patterns that include some features that are similar to those found in adolescents and some features that are distinct in childhood. Furthermore, not all children respond in similar ways to peer stressors. Already by 7 years of age, prior real-life experiences with peers (e.g., peer rejection) as well as personal attributes that increment risk (e.g., social reticence) interact with neural processing of peer stressors. Much remains to be done, with a need for longitudinal multimodal functional and structural imaging research that situates children's school-based peer experiences in a broader social context. Our hope is that this review and suggestions for future research in this rapidly developing field serve to strengthen the empirical literature and inform the development of even more effective prevention and intervention programs designed to alleviate the effects of stress in peer relations in childhood.

References

Achterberg, M., van Duijvenvoorde, A. C. K., van der Meulen, M., Euser, S., Bakermans-Kranenburg, M. J., & Crone, E. A. (2017). The neural and behavioral correlates of social evaluation in childhood. *Developmental Cognitive Neuroscience, 24*, 107–117. https://doi.org/10.1016/j.dcn.2017.02.007

Achterberg, M., van Duijvenvoorde, A. C. K., van der Meulen, M., Bakermans-Kranenburg, M. J., & Crone, E. A. (2018). Heritability of aggression following social evaluation in middle childhood: An fMRI study. *Human Brain Mapping, 39*, 2828–2841. https://doi.org/10.1002/hbm.24043

Albert, D., Chein, J., & Steinberg, L. (2013). The teenage brain: Peer influences on adolescent decision making. *Current Directions in Psychological Science, 22*, 114–120. https://doi.org/10.1177/0963721412471347

Alkire, D., Levitas, D., Warnell, K. R., & Redcay, E. (2018). Social interaction recruits mentalizing and reward systems in middle childhood. *Human Brain Mapping, 39*, 3928–3942. https://doi.org/10.1002/hbm.24221

Andersen, S. L. (2003). Trajectories of brain development: Point of vulnerability or window of opportunity? *Neuroscience & Biobehavioral Reviews, 27*, 3–18. https://doi.org/10.1016/S0149-7634(03)00005-8

Arnsten, A. F. T. (2009). Stress signalling pathways that impair prefrontal cortex structure and function. *Nature Reviews Neuroscience, 10*, 410. https://doi.org/10.1038/nrn2648

Asscheman, J. S., Koot, S., Ma, I., Buil, J. M., Krabbendam, L., Cillessen, A. H. N., & van Lier, P. A. C. (2019). Heightened neural sensitivity to social exclusion in boys with a history of low peer preference during primary school. *Developmental Cognitive Neuroscience, 38*, 100673. https://doi.org/10.1016/j.dcn.2019.100673

Baddam, S., Laws, H., Crawford, J. L., Wu, J., Bolling, D. Z., Mayes, L. C., & Crowley, M. J. (2016). What they bring: Baseline psychological distress differentially predicts neural response in social exclusion by children's friends and strangers in best friend dyads. *Social Cognitive and Affective Neuroscience, 11*, 1729–1740. https://doi.org/10.1093/scan/nsw083

Baldini, S., Restani, L., Baroncelli, L., Coltelli, M., Franco, R., Cenni, M. C., … Berardi, N. (2013). Enriched early life experiences reduce adult anxiety-like behavior in rats: A role for insulin-like growth factor 1. *Journal of Neuroscience, 33*, 11715–11723.

Belsky, J., & Pluess, M. (2009). Beyond diathesis stress: Differential susceptibility to environmental influences. *Psychological Bulletin, 135*, 885.

Berndt, T. J. (2002). Friendship quality and social development. *Current Directions in Psychological Science, 11*, 7–10. https://doi.org/10.1111/1467-8721.00157

Blackhart, G. C., Eckel, L. A., & Tice, D. M. (2007). Salivary cortisol in response to acute social rejection and acceptance by peers. *Biological Psychology, 75*, 267–276. https://doi.org/10.1016/j.biopsycho.2007.03.005

Bolton, J. L., Molet, J., Ivy, A., & Baram, T. Z. (2017). New insights into early-life stress and behavioral outcomes. *Current Opinion in Behavioral Sciences, 14*, 133–139. https://doi.org/10.1016/j.cobeha.2016.12.012

Boyce, W. T., & Ellis, B. J. (2005). Biological sensitivity to context: I. an evolutionary–developmental theory of the origins and functions of stress reactivity. *Development and Psychopathology, 17*, 271–301. https://doi.org/10.1017/S0954579405050145

Branchi, I., D'Andrea, I., Fiore, M., Di Fausto, V., Aloe, L., & Alleva, E. (2006). Early social enrichment shapes social behavior and nerve growth factor and brain-derived neurotrophic factor levels in the adult mouse brain. *Biological Psychiatry, 60*, 690–696. https://doi.org/10.1016/j.biopsych.2006.01.005

Brett, B. E., & de Weerth, C. (2019). The microbiota–gut–brain axis: A promising avenue to foster healthy developmental outcomes. *Developmental Psychobiology, 61*, 772–782. https://doi.org/10.1002/dev.21824

Cacioppo, S., Frum, C., Asp, E., Weiss, R. M., Lewis, J. W., & Cacioppo, J. T. (2013). A quantitative meta-analysis of functional imaging studies of social rejection. *Scientific Reports, 3*, 2027. https://doi.org/10.1038/srep02027

Callaghan, B. L., & Tottenham, N. (2016). The stress acceleration hypothesis: Effects of early-life adversity on emotion circuits and behavior. *Current Opinion in Behavioral Sciences, 7*, 76–81. https://doi.org/10.1016/j.cobeha.2015.11.018

Chein, J., Albert, D., O'Brien, L., Uckert, K., & Steinberg, L. (2011). Peers increase adolescent risk taking by enhancing activity in the brain's reward circuitry. *Developmental Science, 14*, F1–10. https://doi.org/10.1111/j.1467-7687.2010.01035.x

Chen, Y., & Baram, T. Z. (2015). Toward understanding how early-life stress reprograms cognitive and emotional brain networks. *Neuropsychopharmacology, 41*, 197. https://doi.org/10.1038/npp.2015.181

Cicchetti, D. (2002). The impact of social experience on neurobiological systems: Illustration from a constructivist view of child maltreatment. *Cognitive Development, 17*, 1407–1428. https://doi.org/10.1016/S0885-2014(02)00121-1

Cohen, S., Gottlieb, B. H., & Underwood, L. G. (2000). Social relationships and health. In S. Cohen, L. G. Underwood, & B. H. Gottlieb (Eds.), *Social support measurement and intervention: A guide for health and social scientists* (pp. 3–25). Oxford University Press.

Collins, S. M., Surette, M., & Bercik, P. (2012). The interplay between the intestinal microbiota and the brain. *Nature Reviews Microbiology, 10*, 735. https://doi.org/10.1038/nrmicro2876

Copeland, W. E., Wolke, D., Angold, A., & Costello, E. J. (2013). Adult psychiatric outcomes of bullying and being bullied by peers in childhood and adolescence. *JAMA Psychiatry, 70*, 419–426. https://doi.org/10.1001/jamapsychiatry.2013.504

Coplan, R. J., & Arbeau, K. A. (2009). Peer interactions and play in early childhood. In K. H. Rubin, W. M. Bukowski, & B. Laursen (Eds.), *Handbook of peer interactions, relationships, and groups*. The Guilford Press.

Crowley, M. J., Wu, J., Molfese, P. J., & Mayes, L. C. (2010). Social exclusion in middle childhood: Rejection events, slow-wave neural activity, and ostracism distress. *Social Neuroscience, 5*, 483–495. https://doi.org/10.1080/17470919.2010.500169

Curley, J. P., & Champagne, F. A. (2016). Influence of maternal care on the developing brain: Mechanisms, temporal dynamics and sensitive periods. *Frontiers in Neuroendocrinology, 40*, 52–66. https://doi.org/10.1016/j.yfrne.2015.11.001

Dalgleish, T., Walsh, N. D., Mobbs, D., Schweizer, S., van Harmelen, A.-L., Dunn, B., … Stretton, J. (2017). Social pain and social gain in the adolescent brain: A common neural circuitry underlying both positive and negative social evaluation. *Scientific Reports, 7*, 42010. https://doi.org/10.1038/srep42010

Davey, C. G., Allen, N. B., Harrison, B. J., Dwyer, D. B., & Yücel, M. (2010). Being liked activates primary reward and midline self-related brain regions. *Human Brain Mapping, 31*, 660–668. https://doi.org/10.1002/hbm.20895

de Kloet, E. R. (2003). Hormones, brain and stress. *Endocrine Regulations, 37*, 51.

de Kloet, E. R., Joëls, M., & Holsboer, F. (2005). Stress and the brain: From adaptation to disease. *Nature Reviews Neuroscience, 6*, 463–475. https://doi.org/10.1038/nrn1683

Delgado, M. R. (2007). Reward-related responses in the human striatum. In *Reward and decision making in corticobasal ganglia networks* (Vol. 1104, pp. 70–88). Blackwell Publishing.

Dickerson, S. S., & Zoccola, P. M. (2013). Cortisol responses to social exclusion. In C. N. DeWall (Ed.), *The Oxford handbook of social exclusion* (pp. 143–151). Oxford University Press.

Dippel, G., & Beste, C. (2015). A causal role of the right inferior frontal cortex in implementing strategies for multi-component behaviour. *Nature Communications, 6*, 6587. https://doi.org/10.1038/ncomms7587

Dodge, K. A., Lansford, J. E., Burks, V. S., Bates, J. E., Pettit, G. S., Fontaine, R., & Price, J. M. (2003). Peer rejection and social information-processing factors in the development of aggressive behavior problems in children. *Child Development, 74*, 374–393. https://doi.org/10.1111/1467-8624.7402004

du Plessis, M. R., Smeekens, S., Cillessen, A. H. N., Whittle, S., & Güroğlu, B. (2019). Bullying the brain? Longitudinal links between childhood peer victimization, cortisol, and adolescent brain structure. *Frontiers in Psychology, 9*, 2706–2706. https://doi.org/10.3389/fpsyg.2018.02706

Eisenberger, N. I. (2012). The pain of social disconnection: Examining the shared neural underpinnings of physical and social pain. *Nature Reviews Neuroscience, 13*, 421. https://doi.org/10.1038/nrn3231

Eisenberger, N. I., Lieberman, M. D., & Williams, K. D. (2003). Does rejection hurt? An fMRI study of social exclusion. *Science, 302*, 290–292. https://doi.org/10.1126/science.1089134

Endedijk, H. M., Meyer, M., Bekkering, H., Cillessen, A. H. N., & Hunnius, S. (2017). Neural mirroring and social interaction: Motor system involvement during action observation relates to early peer cooperation. *Developmental Cognitive Neuroscience, 24*, 33–41. https://doi.org/10.1016/j.dcn.2017.01.001

Feldman, D. E., & Knudsen, E. I. (1998). Experience-dependent plasticity and the maturation of glutamatergic synapses. *Neuron, 20*, 1067–1071.

Ferreri, L., Bigand, E., Perrey, S., & Bugaiska, A. (2014). The promise of near-infrared spectroscopy (NIRS) for psychological research: A brief review. *L'Annee Psychologique, 114*, 537–569.

Fox, S. E., Levitt, P., & Nelson, C. A., III. (2010). How the timing and quality of early experiences influence the development of brain architecture. *Child Development, 81*, 28–40. https://doi.org/10.1111/j.1467-8624.2009.01380.x

Gardner, M., & Steinberg, L. (2005). Peer influence on risk taking, risk preference, and risky decision making in adolescence and adulthood: An experimental study. *Developmental Psychology, 41*, 625–635. https://doi.org/10.1037/0012-1649.41.4.625

Ge, X., & Natsuaki, M. N. (2009). In search of explanations for early pubertal timing effects on developmental psychopathology. *Current Directions in Psychological Science, 18*, 327–331. https://doi.org/10.1111/j.1467-8721.2009.01661.x

Giedd, J. N., Blumenthal, J., Jeffries, N. O., Castellanos, F. X., Liu, H., Zijdenbos, A., … Rapoport, J. L. (1999). Brain development during childhood and adolescence: A longitudinal MRI study. *Nature Neuroscience, 2*, 861–863. https://doi.org/10.1038/13158

Gifford-Smith, M. E., & Brownell, C. A. (2003). Childhood peer relationships: Social acceptance, friendships, and peer networks. *Journal of School Psychology, 41*, 235–284. https://doi.org/10.1016/S0022-4405(03)00048-7

Goes, T. C., Antunes, F. D., & Teixeira-Silva, F. (2015). Environmental enrichment for adult rats: Effects on trait and state anxiety. *Neuroscience Letters, 584*, 93–96. https://doi.org/10.1016/j.neulet.2014.10.004

Gogtay, N., Giedd, J. N., Lusk, L., Hayashi, K. M., Greenstein, D., Vaituzis, A. C., … Toga, A. W. (2004). Dynamic mapping of human cortical development during childhood through early adulthood. *Proceedings of the National Academy of Sciences, 101*, 8174–8179.

Greenberg, M. T., Domitrovich, C. E., Weissberg, R. P., & Durlak, J. A. (2017). Social and emotional learning as a public health approach to education. *The Future of Children, 27*, 13–32.

Gregory, A. M., Light-Häusermann, J. H., Rijsdijk, F., & Eley, T. C. (2009). Behavioral genetic analyses of prosocial behavior in adolescents. *Developmental Science, 12*, 165–174. https://doi.org/10.1111/j.1467-7687.2008.00739.x

Gunther Moor, B., van Leijenhorst, L., Rombouts, S. A., Crone, E. A., & Van der Molen, M. W. (2010). Do you like me? Neural correlates of social evaluation and developmental trajectories. *Social Neuroscience, 5*, 461–482. https://doi.org/10.1080/17470910903526155

Guyer, A. E., & Jarcho, J. M. (2018). Neuroscience and peer relations. In W. M. Bukowski, B. Laursen, & K. H. Rubin (Eds.), *Handbook of peer interactions, relationships, and groups* (2nd ed.). The Guilford Press.

Guyer, A. E., McClure-Tone, E. B., Shiffrin, N. D., Pine, D. S., & Nelson, E. E. (2009). Probing the neural correlates of anticipated peer evaluation in adolescence. *Child Development, 80*, 1000–1015. https://doi.org/10.1111/j.1467-8624.2009.01313.x

Guyer, A. E., Choate, V. R., Pine, D. S., & Nelson, E. E. (2011). Neural circuitry underlying affective response to peer feedback in adolescence. *Social Cognitive and Affective Neuroscience, 7*, 81–92. https://doi.org/10.1093/scan/nsr043

Hartup, W. W. (1996). The company they keep: Friendships and their developmental significance. *Child Development, 67*, 1–13. https://doi.org/10.1111/j.1467-8624.1996.tb01714.x

Hebb, D. O. (1949). The organization of behavior: A neuropsycholocigal theory. In *A wiley book in clinical psychology* (pp. 62–78). Wiley.

Heise, D. R. (1969). Separating reliability and stability in test-retest correlation. *American Sociological Review, 34*, 93–101. https://doi.org/10.2307/2092790

Herting, M. M., Gautam, P., Chen, Z., Mezher, A., & Vetter, N. C. (2018). Test-retest reliability of longitudinal task-based fMRI: Implications for developmental studies. *Developmental Cognitive Neuroscience, 33*, 17–26. https://doi.org/10.1016/j.dcn.2017.07.001

Hoffmann, F., Puetz, V. B., Viding, E., Sethi, A., Palmer, A., & McCrory, E. J. (2017). Risk-taking, peer-influence and child maltreatment: A neurocognitive investigation. *Social Cognitive and Affective Neuroscience, 13*, 124–134. https://doi.org/10.1093/scan/nsx124

Holmes, C. J., Kim-Spoon, J., & Deater-Deckard, K. (2016). Linking executive function and peer problems from early childhood through middle adolescence. *Journal of Abnormal Child Psychology, 44*, 31–42. https://doi.org/10.1007/s10802-015-0044-5

Holtmaat, A., & Svoboda, K. (2009). Experience-dependent structural synaptic plasticity in the mammalian brain. *Nature Reviews Neuroscience, 10*, 647. https://doi.org/10.1038/nrn2699

Janak, P. H., & Tye, K. M. (2015). From circuits to behaviour in the amygdala. *Nature, 517*, 284. https://doi.org/10.1038/nature14188

Jarcho, J. M., Leibenluft, E., Walker, O. L., Fox, N. A., Pine, D. S., & Nelson, E. E. (2013). Neuroimaging studies of pediatric social anxiety: Paradigms, pitfalls and a new direction for investigating the neural mechanisms. *Biology of Mood & Anxiety Disorders, 3*, 14. https://doi.org/10.1186/2045-5380-3-14

Jarcho, J. M., Davis, M. M., Shechner, T., Degnan, K. A., Henderson, H. A., Stoddard, J., … Nelson, E. E. (2016). Early-childhood social reticence predicts brain function in preadolescent youths during distinct forms of peer evaluation. *Psychological Science, 27*, 821–835. https://doi.org/10.1177/0956797616638319

Jarcho, J. M., Grossman, H. Y., Guyer, A. E., Quarmley, M., Smith, A. R., Fox, N. A., … Nelson, E. E. (2019). Connecting childhood wariness to adolescent social anxiety through the brain and peer experiences. *Journal of Abnormal Child Psychology*. https://doi.org/10.1007/s10802-019-00543-4

Karatsoreos, I. N., & McEwen, B. S. (2013). Annual research review: The neurobiology and physiology of resilience and adaptation across the life course. *Journal of Child Psychology and Psychiatry, 54*, 337–347. https://doi.org/10.1111/jcpp.12054

Karremans, J. C., Heslenfeld, D. J., van Dillen, L. F., & Van Lange, P. A. M. (2011). Secure attachment partners attenuate neural responses to social exclusion: An fMRI investigation. *International Journal of Psychophysiology, 81*, 44–50. https://doi.org/10.1016/j.ijpsycho.2011.04.003

Kelley, N. J., Hortensius, R., Schutter, D. J. L. G., & Harmon-Jones, E. (2017). The relationship of approach/avoidance motivation and asymmetric frontal cortical activity: A review of studies manipulating frontal asymmetry. *International Journal of Psychophysiology, 119*, 19–30. https://doi.org/10.1016/j.ijpsycho.2017.03.001

Kempermann, G. (2019). Environmental enrichment, new neurons and the neurobiology of individuality. *Nature Reviews Neuroscience, 20*, 235–245. https://doi.org/10.1038/s41583-019-0120-x

Kennedy, D. P., & Adolphs, R. (2012). The social brain in psychiatric and neurological disorders. *Trends in Cognitive Sciences, 16*, 559–572. https://doi.org/10.1016/j.tics.2012.09.006

Kilford, E. J., Garrett, E., & Blakemore, S.-J. (2016). The development of social cognition in adolescence: An integrated perspective. *Neuroscience & Biobehavioral Reviews, 70*, 106–120. https://doi.org/10.1016/j.neubiorev.2016.08.016

Knafo, A., & Plomin, R. (2006). Prosocial behavior from early to middle childhood: Genetic and environmental influences on stability and change. *Developmental Psychology, 42*, 771.

Kok, R., Thijssen, S., Bakermans-Kranenburg, M. J., Jaddoe, V. W. V., Verhulst, F. C., White, T., … Tiemeier, H. (2015). Normal variation in early parental sensitivity predicts child structural brain development. *Journal of the American Academy of Child & Adolescent Psychiatry, 54*, 824-831.e821. https://doi.org/10.1016/j.jaac.2015.07.009

Kok, R., Prinzie, P., Bakermans-Kranenburg, M. J., Verhulst, F. C., White, T., Tiemeier, H., & van Ijzendoorn, M. H. (2018). Socialization of prosocial behavior: Gender differences in the mediating role of child brain volume. *Child Neuropsychology, 24*, 723–733. https://doi.org/10.1080/09297049.2017.1338340

Kopala-Sibley, D. C., Cyr, M., Finsaas, M. C., Orawe, J., Huang, A., Tottenham, N., & Klein, D. N. (2018). Early childhood parenting predicts late childhood brain functional connectivity during emotion perception and reward processing. *Child Development*. https://doi.org/10.1111/cdev.13126

Lahat, A., Walker, O. L., Lamm, C., Degnan, K. A., Henderson, H. A., & Fox, N. A. (2014). Cognitive conflict links behavioural inhibition and social problem solving during social exclusion in childhood. *Infant and Child Development, 23*, 273–282. https://doi.org/10.1002/icd.1845

Lansford, J. E., Malone, P. S., Dodge, K. A., Pettit, G. S., & Bates, J. E. (2010). Developmental cascades of peer rejection, social information processing biases, and aggression during middle childhood. *Development and Psychopathology, 22*, 593–602. https://doi.org/10.1017/S0954579410000301

Lebel, C., & Beaulieu, C. (2011). Longitudinal development of human brain wiring continues from childhood into adulthood. *The Journal of Neuroscience, 31*, 10937–10947. https://doi.org/10.1523/jneurosci.5302-10.2011

Lee, H. S., Lee, J. E., Lee, K. U., & Kim, Y. H. (2014). Neural changes associated with emotion processing in children experiencing peer rejection: A functional MRI study. *Journal of Korean Medical Science, 29*, 1293–1300.

Lehmann, M. L., & Herkenham, M. (2011). Environmental enrichment confers stress resiliency to social defeat through an infralimbic cortex-dependent neuroanatomical pathway. *The Journal of Neuroscience, 31*, 6159. https://doi.org/10.1523/JNEUROSCI.0577-11.2011

Lejuez, C. W., Read, J. P., Kahler, C. W., Richards, J. B., Ramsey, S. E., Stuart, G. L., … Brown, R. A. (2002). Evaluation of a behavioral measure of risk taking: The balloon analogue risk task (BART). *Journal of Experimental Psychology: Applied, 8*, 75–84. https://doi.org/10.1037/1076-898X.8.2.75

Lloyd-Fox, S., Blasi, A., & Elwell, C. E. (2010). Illuminating the developing brain: The past, present and future of functional near infrared spectroscopy. *Neuroscience & Biobehavioral Reviews, 34*, 269–284. https://doi.org/10.1016/j.neubiorev.2009.07.008

London, B., Downey, G., Bonica, C., & Paltin, I. (2007). Social causes and consequences of rejection sensitivity. *Journal of Research on Adolescence, 17*, 481–506. https://doi.org/10.1111/j.1532-7795.2007.00531.x

Lucassen, P., Korosi, A., Krugers, H., & Oomen, C. (2016). Early life stress-and sex-dependent effects on hippocampal neurogenesis. In G. Fink (Ed.), *Stress: Neuroendocrinology and neurobiology: Handbook of stress series* (Vol. 2, p. 135). Elsevier Inc.

Lupien, S. J., McEwen, B. S., Gunnar, M. R., & Heim, C. (2009). Effects of stress throughout the lifespan on the brain, behaviour and cognition. *Nature Reviews Neuroscience, 10*, 434. https://doi.org/10.1038/nrn2639

Masten, C. L., Eisenberger, N. I., Borofsky, L. A., Pfeifer, J. H., McNealy, K., Mazziotta, J. C., & Dapretto, M. (2009). Neural correlates of social exclusion during adolescence: Understanding the distress of peer rejection. *Social Cognitive and Affective Neuroscience, 4*, 143–157. https://doi.org/10.1093/scan/nsp007

McDonald, K. L., & Asher, S. R. (2018). Peer acceptance, peer rejection, and popularity: Social-cognitive and behavioral perspectives. In W. M. Bukowski, B. Laursen, & K. H. Rubin (Eds.), *Handbook of peer interactions, relationships, and groups*. The Guilford Press.

McHugh, S. B., Barkus, C., Huber, A., Capitão, L., Lima, J., Lowry, J. P., & Bannerman, D. M. (2014). Aversive prediction error signals in the amygdala. *The Journal of Neuroscience, 34*, 9024. https://doi.org/10.1523/JNEUROSCI.4465-13.2014

Mills, K. L., Lalonde, F., Clasen, L. S., Giedd, J. N., & Blakemore, S. J. (2014). Developmental changes in the structure of the social brain in late childhood and adolescence. *Social Cognitive Affective Neuroscience, 9*, 123–131. https://doi.org/10.1093/scan/nss113

Mills, K. L., Goddings, A.-L., Herting, M. M., Meuwese, R., Blakemore, S.-J., Crone, E. A., ... Tamnes, C. K. (2016). Structural brain development between childhood and adulthood: Convergence across four longitudinal samples. *NeuroImage, 141*, 273–281. https://doi.org/10.1016/j.neuroimage.2016.07.044

Nelson, E. E., Leibenluft, E., McClure, E. B., & Pine, D. S. (2005). The social re-orientation of adolescence: A neuroscience perspective on the process and its relation to psychopathology. *Psychological Medicine, 35*, 163–174. https://doi.org/10.1017/s0033291704003915

O'Mahony, S. M., Clarke, G., Dinan, T. G., & Cryan, J. F. (2017). Early-life adversity and brain development: Is the microbiome a missing piece of the puzzle? *Neuroscience, 342*, 37–54. https://doi.org/10.1016/j.neuroscience.2015.09.068

Ouellet-Morin, I., Danese, A., Bowes, L., Shakoor, S., Ambler, A., Pariante, C. M., ... Arseneault, L. (2011). A discordant monozygotic twin design shows blunted cortisol reactivity among bullied children. *Journal of the American Academy of Child & Adolescent Psychiatry, 50*, 574-582.e573. https://doi.org/10.1016/j.jaac.2011.02.015

Parker, J. G., & Asher, S. R. (1987). Peer relations and later personal adjustment: Are low-accepted children at risk? *Psychological Bulletin, 102*, 357.

Peper, J. S., Brouwer, R. M., Boomsma, D. I., Kahn, R. S., & Hulshoff Pol, H. E. (2007). Genetic influences on human brain structure: A review of brain imaging studies in twins. *Human Brain Mapping, 28*, 464–473. https://doi.org/10.1002/hbm.20398

Perini, I., Gustafsson, P. A., Hamilton, J. P., Kämpe, R., Zetterqvist, M., & Heilig, M. (2018). The salience of self, not social pain, is encoded by dorsal anterior cingulate and insula. *Scientific Reports, 8*, 6165. https://doi.org/10.1038/s41598-018-24658-8

Peters, E., Riksen-Walraven, J. M., Cillessen, A. H. N., & de Weerth, C. (2011). Peer rejection and HPA activity in middle childhood: Friendship makes a difference. *Child Development, 82*, 1906–1920. https://doi.org/10.1111/j.1467-8624.2011.01647.x

Pfeifer, J. H., Lieberman, M. D., & Dapretto, M. (2007). "I know you are but what am I?!": Neural bases of self- and social knowledge retrieval in children and adults. *Journal of Cognitive Neuroscience, 19*, 1323–1337. https://doi.org/10.1162/jocn.2007.19.8.1323

Riem, M. M. E., Bakermans-Kranenburg, M. J., Huffmeijer, R., & van Ijzendoorn, M. H. (2013). Does intranasal oxytocin promote prosocial behavior to an excluded fellow player? A randomized-controlled trial with Cyberball. *Psychoneuroendocrinology, 38*, 1418–1425. https://doi.org/10.1016/j.psyneuen.2012.12.023

Rogers, G. B., Keating, D. J., Young, R. L., Wong, M. L., Licinio, J., & Wesselingh, S. (2016). From gut dysbiosis to altered brain function and mental illness: Mechanisms and pathways. *Molecular Psychiatry, 21*, 738. https://doi.org/10.1038/mp.2016.50

Rubin, K. H., Bukowski, W. M., & Parker, J. G. (2006). Peer interactions, relationships, and groups. In *Handbook of child psychology* (Vol. III). Wiley.

Rudolph, K. D., Miernicki, M. E., Troop-Gordon, W., Davis, M. M., & Telzer, E. H. (2016). Adding insult to injury: Neural sensitivity to social exclusion is associated with internalizing symptoms in chronically peer-victimized girls. *Social Cognitive and Affective Neuroscience, 11*, 829–842. https://doi.org/10.1093/scan/nsw021

Sartori, L., Betti, S., & Castiello, U. (2013). When mirroring is not enough: That is, when only a complementary action will do (the trick). *Neuroreport, 24*, 601–604. https://doi.org/10.1097/WNR.0b013e3283630a66

Saylor, C. F., Williams, K. D., Nida, S. A., McKenna, M. E., Twomey, K. E., & Macias, M. M. (2013). Ostracism in pediatric populations: Review of theory and research. *Journal of Developmental & Behavioral Pediatrics, 34*, 279–287.

Schneider, B. H., Woodburn, S., del Toro, M. d. P. S., & Udvari, S. J. (2005). Cultural and gender differences in the implications of competition for early adolescent friendship. *Merrill-Palmer Quarterly, 51*, 163–191.

Schriber, R. A., & Guyer, A. E. (2016). Adolescent neurobiological susceptibility to social context. *Developmental Cognitive Neuroscience, 19*, 1–18. https://doi.org/10.1016/j.dcn.2015.12.009

Shaw, P., Eckstrand, K., Sharp, W., Blumenthal, J., Lerch, J. P., Greenstein, D., ... Rapoport, J. L. (2007). Attention-deficit/hyperactivity disorder is characterized by a delay in cortical maturation. *Proceedings of the National Academy of Sciences, 104*, 19649. https://doi.org/10.1073/pnas.0707741104

Shaw, P., Gogtay, N., & Rapoport, J. (2010). Childhood psychiatric disorders as anomalies in neurodevelopmental trajectories. *Human Brain Mapping, 31*, 917–925. https://doi.org/10.1002/hbm.21028

Shenhav, A., Botvinick, M. M., & Cohen, J. D. (2013). The expected value of control: An integrative theory of anterior cingulate cortex function. *Neuron, 79*, 217–240. https://doi.org/10.1016/j.neuron.2013.07.007

Shonkoff, J. P., & Levitt, P. (2010). Neuroscience and the future of early childhood policy: Moving from why to what and how. *Neuron, 67*, 689–691. https://doi.org/10.1016/j.neuron.2010.08.032

Silston, B., Bassett, D. S., & Mobbs, D. (2018). How dynamic brain networks tune social behavior in real time. *Current Directions in Psychological Science, 27*, 413–421. https://doi.org/10.1177/0963721418773362

Somerville, L. H. (2013). The teenage brain: Sensitivity to social evaluation. *Current Directions in Psychological Science, 22*, 121–127. https://doi.org/10.1177/0963721413476512

Somerville, L. H., Heatherton, T. F., & Kelley, W. M. (2006). Anterior cingulate cortex responds differentially to expectancy violation and social rejection. *Nature Neuroscience, 9*, 1007–1008. https://doi.org/10.1038/nn1728

Sparling, J. E., Baker, S. L., & Bielajew, C. (2018). Effects of combined pre- and post-natal enrichment on anxiety-like, social, and cognitive behaviours in juvenile and adult rat offspring. *Behavioural Brain Research, 353*, 40–50. https://doi.org/10.1016/j.bbr.2018.06.033

Stanley, D. A., & Adolphs, R. (2013). Toward a neural basis for social behavior. *Neuron, 80*, 816–826. https://doi.org/10.1016/j.neuron.2013.10.038

Tau, G. Z., & Peterson, B. S. (2009). Normal development of brain circuits. *Neuropsychopharmacology, 35*, 147. https://doi.org/10.1038/npp.2009.115

Taylor, S. E. (2011). Tend and befriend theory. In P. van Lange, A. Kruglanski, & E. T. Higgins (Eds.), *Handbook of theories of social psychology* (p. 32). Sage Publications.

Telzer, E. H., Fuligni, A. J., Lieberman, M. D., Miernicki, M. E., & Galvan, A. (2015). The quality of adolescents' peer relationships modulates neural sensitivity to risk taking. *Social Cognitive and Affective Neuroscience, 10*, 389–398. https://doi.org/10.1093/scan/nsu064

Telzer, E. H., McCormick, E. M., Peters, S., Cosme, D., Pfeifer, J. H., & van Duijvenvoorde, A. C. K. (2018). Methodological considerations for developmental longitudinal fMRI research. *Developmental Cognitive Neuroscience, 33*, 149–160. https://doi.org/10.1016/j.dcn.2018.02.004

van der Meulen, M., Steinbeis, N., Achterberg, M., Bilo, E., van den Bulk, B. G., & van, I. M. H., & Crone, E. A. (2017). The neural correlates of dealing with social exclusion in childhood. *Neuropsychologia*. https://doi.org/10.1016/j.neuropsychologia.2017.07.008

van der Meulen, M., Steinbeis, N., Achterberg, M., van Ijzendoorn, M. H., & Crone, E. A. (2018). Heritability of neural reactions to social exclusion and prosocial compensation in middle childhood. *Developmental Cognitive Neuroscience, 34*, 42–52. https://doi.org/10.1016/j.dcn.2018.05.010

van Noordt, S. J., White, L. O., Wu, J., Mayes, L. C., & Crowley, M. J. (2015). Social exclusion modulates event-related frontal theta and tracks ostracism distress in children. *NeuroImage, 118*, 248–255. https://doi.org/10.1016/j.neuroimage.2015.05.085

van Praag, H., Kempermann, G., & Gage, F. H. (2000). Neural consequences of enviromental enrichment. *Nature Reviews Neuroscience, 1*, 191–198. https://doi.org/10.1038/35044558

van Wijk, I. C., van den Bulk, B. G., Euser, S., Bakermans-Kranenburg, M. J., van Ijzendoorn, M. H., & Huffmeijer, R. (2019). Social judgments, frontal asymmetry, and aggressive behavior in young children: A replication study using EEG. *Neuropsychologia, 126*, 120–127. https://doi.org/10.1016/j.neuropsychologia.2017.06.022

VanTieghem, M. R., & Tottenham, N. (2018). Neurobiological programming of early life stress: Functional development of amygdala-prefrontal circuitry and vulnerability for stress-related psychopathology. In E. Vermetten, D. G. Baker, & V. B. Risbrough (Eds.), *Behavioral neurobiology of PTSD* (pp. 117–136). Springer International Publishing.

Vijayakumar, N., Cheng, T. W., & Pfeifer, J. H. (2017). Neural correlates of social exclusion across ages: A coordinate-based meta-analysis of functional MRI studies. *NeuroImage, 153*, 359–368. https://doi.org/10.1016/j.neuroimage.2017.02.050

Wachs, T. D., Georgieff, M., Cusick, S., & McEwen, B. S. (2014). Issues in the timing of integrated early interventions: Contributions from nutrition, neuroscience, and psychological research. *Annals of the New York Academy of Sciences, 1308*, 89–106. https://doi.org/10.1111/nyas.12314

Wang, Y., & Olson, I. R. (2018). The original social network: White matter and social cognition. *Trends in Cognitive Sciences, 22*, 504–516. https://doi.org/10.1016/j.tics.2018.03.005

Wang, H., Braun, C., & Enck, P. (2017). How the brain reacts to social stress (exclusion) – A scoping review. *Neuroscience & Biobehavioral Reviews, 80*, 80–88. https://doi.org/10.1016/j.neubiorev.2017.05.012

Wang, S., Harvey, L., Martin, R., van der Beek, E. M., Knol, J., Cryan, J. F., & Renes, I. B. (2018). Targeting the gut microbiota to influence brain development and function in early life. *Neuroscience & Biobehavioral Reviews, 95*, 191–201. https://doi.org/10.1016/j.neubiorev.2018.09.002

Wang, Q., Zhang, H., Wee, C.-Y., Lee, A., Poh, J. S., Chong, Y.-S., … Qiu, A. (2019). Maternal sensitivity predicts anterior hippocampal functional networks in early childhood. *Brain Structure and Function, 224*, 1885–1895. https://doi.org/10.1007/s00429-019-01882-0

Warnell, K. R., Sadikova, E., & Redcay, E. (2018). Let's chat: Developmental neural bases of social motivation during real-time peer interaction. *Developmental Science, 21*, e12581. https://doi.org/10.1111/desc.12581

Westenberg, M. P., Drewes, M. J., Goedhart, A. W., Siebelink, B. M., & Treffers, P. D. A. (2004). A developmental analysis of self-reported fears in late childhood through mid-adolescence: Social-evaluative fears on the rise? *Journal of Child Psychology and Psychiatry, 45*, 481–495. https://doi.org/10.1111/j.1469-7610.2004.00239.x

Whitfield-Gabrieli, S., Ghosh, S. S., Nieto-Castanon, A., Saygin, Z., Doehrmann, O., Chai, X. J., … Gabrieli, J. D. E. (2015). Brain connectomics predict response to treatment in social anxiety disorder. *Molecular Psychiatry, 21*, 680. https://doi.org/10.1038/mp.2015.109

Whittle, S., Simmons, J. G., Dennison, M., Vijayakumar, N., Schwartz, O., Yap, M. B. H., … Allen, N. B. (2014). Positive parenting predicts the development of adolescent brain structure: A longitudinal study. *Developmental Cognitive Neuroscience, 8*, 7–17. https://doi.org/10.1016/j.dcn.2013.10.006

Wilcox, T., & Biondi, M. (2015). fNIRS in the developmental sciences. *Wiley Interdisciplinary Reviews: Cognitive Science, 6*, 263–283. https://doi.org/10.1002/wcs.1343

Will, G. J., van Lier, P. A., Crone, E. A., & Güroğlu, B. (2016). Chronic childhood peer rejection is associated with heightened neural responses to social exclusion during adolescence. *Journal of Abnormal Child Psychology, 44*, 43–55. https://doi.org/10.1007/s10802-015-9983-0

Williams, K. D., & Jarvis, B. (2006). Cyberball: A program for use in research on interpersonal ostracism and acceptance. *Behavior Research Methods, 38*, 174–180. https://doi.org/10.3758/BF03192765

Zijlmans, M. A. C., Korpela, K., Riksen-Walraven, J. M., de Vos, W. M., & de Weerth, C. (2015). Maternal prenatal stress is associated with the infant intestinal microbiota. *Psychoneuroendocrinology, 53*, 233–245. https://doi.org/10.1016/j.psyneuen.2015.01.006

Epilogue: Concluding Commentary

Kirby Deater-Deckard and Pol A. C. van Lier

The evidence from correlational and experimental studies is clear: supportive, accepting peer and teacher relationships promote adaptive skills and outcomes, but harsh, rejecting social experiences in school consistently contribute to growth in behavioral, emotional, and academic problems. As a collection, the chapters in the current volume offer key insights into the known, likely, or potential causal processes that not only help explain why these patterns emerge and hold true for long periods of time in development but that identify targets for prevention and intervention within and outside of school settings. In this closing commentary, after briefly summarizing key points, we offer three recommendations for future research that are informed by and build upon the gaps and needs for future research highlighted by the volume authors.

We have organized the chapters with the following framework: structures and processes of peer and teacher social experiences in school settings define the ecological boundary for the current volume and the work we have presented. The various levels and dimensions of this school-based social ecology are thought to directly influence the range of positive, neutral, and negative dyadic and group interactions and processes that together represent the experienced environment at school. These experiences, in turn, can contribute to short-term as well as long-lasting changes in neurobiological structures and processes spanning the genome, the brain and autonomic nervous system, and social cognitive functions that influence decision-making. Furthermore, the relevant individual, dyadic, and group features and dynamics moderate the processes connecting experiences to neurobiological and

K. Deater-Deckard (✉)
Department of Psychological & Brain Sciences, University of Massachusetts Amherst, Amherst, MA, USA
e-mail: kdeaterdeck@umass.edu

P. A. C. van Lier
Clinical Developmental Psychology, Vrije Universiteit Amsterdam, Amsterdam, The Netherlands

P. A. C. van Lier, K. Deater-Deckard (eds.), *Biosocial Interplay During Elementary School*, https://doi.org/10.1007/978-3-031-07109-6_9

cognitive changes. These changes contribute to developmental shifts in a host of normative and maladaptive outcomes ranging from academics to behavior to social-emotional. Lastly, these "outcomes" themselves influence stability or change in peer and teacher relationships and interactions, reflecting bidirectional influences between the developing child's attributes and those of peers and teachers.

Chapter "Prologue: Introduction" identifies four levels of influence in school settings (individual, dyadic interactions, dyadic relationships, and group dynamics) that work together to explain and predict children's developmental trajectories in academic, social, and social-emotional well-being domains. This organizational framework defines the multiple components of the school-based social behavioral system. Chapter "Elementary School Social Experiences with Peers and Teachers: Manifestation and Development" builds and elaborates on this foundation, emphasizing the need to better articulate the independent, additive, and interactive effects that operate over time (measured longitudinally) and that are mediated by not only behavior but cognitive and affective features of experience and learning. In order to understand the promotive and risk-inducing effects of particular experiences and characteristics (e.g., aggression, rejection, prosocial acts), future research will benefit by integrating information across multiple levels and across both peer and teacher relationships.

For incorporating the genome, chapter "How Peers and Teachers Shape Elementary School Children's Academic and Socioemotional Development" highlights the gene-environment correlation and interaction effects through which genetic differences between children transact with school social experiences, to account for the wide range of developmental outcome trajectories that are observed. In addition, as noted in chapter "How Peers and Teachers Shape Elementary School Children's Academic and Socioemotional Development" and described in detail in chapter "School Social Relations and Child Development: Gene-Environment Interplay", the nascent literature on epigenetic differences due to adverse childhood experiences in humans is suggestive but mixed. It is clear that there will be a need to incorporate direct measurement of epigenetic information (as well as other indicators of modifications to gene products in the body), to more fully understand the functional differences in gene expression that can explain predictive effects arising from gene-environment interactions involving adverse social environments in schools and other contexts.

Moving upward in the internal system of the body from the genome, chapter "The Impact of School Social Experiences on Socioemotional and Behavioral Problems: The Hypothesized Role of DNA Methylation" integrates a large and growing literature on the impact of aversive social experiences with caregivers and peers, on constantly developing stress response and regulatory systems. Building on this foundation, chapter "Biological Embedding of Peer Experiences: The Contribution of Peer Adversity to Stress Regulation" presents a framework of testing competing hypotheses about how a dysregulated stress response can interfere with decision-making processes—alterations in social cognition that have powerful consequences for children's social relationships and functioning. The concluding chapter "School Social Relations, Self-Regulation, and Social Decision-Making" describes theory and empirical evidence regarding links between aversive peer social interactions and experiences with shifts in neural activity in the brain, reflecting adaptations that influence sensitivity and responsiveness to social cues. When scanning across the levels of the intertwined systems of stress response and self-regulation, it is clear that there are many challenges to operationalizing and measuring the major components of these systems. Future research will require a diversification of measurement of inputs, mediators, and outcomes before the field will be able to rigorously test competing theories of how it is that such experiences become embedded within neurobiological processes and are manifested in lasting alterations to neural and social cognitive decision-making.

Recommendations

Based on the body of theories and evidence presented in this volume, and upon reflecting on many of the key points from the chapters' authors regarding future research needs, we offer three recommendations to integrate what we view as critically important next steps.

Recommendation 1: Study Virtual and In-Person School Social Experiences

Our first recommendation reflects a timely situation at the time we were completing this book. In 2020 and continuing through 2022, the COVID-19 pandemic caused rapid and lasting changes in how families and schools functioned. As a result and as school systems around the world shifted to remote learning (World Bank, 2021), interest in virtual classrooms as a learning environment has accelerated dramatically. This has included rapid growth concerns about the positive and negative social experiences that school-aged youth experience online—a line of research that began about two decades ago with the advent of cellular- and Internet-based modes of communication (Keith & Martin, 2005; Twenge et al., 2018). One lesson from this for scholars, scientists, and policymakers is that there has never been a more urgent need to fully incorporate all venues (face to face and virtual) in which children live their school- and nonschool-based social lives.

Although it is too soon to tell how long-lasting pandemic-related changes in children's school time will be, many educational policymakers and practitioners are anticipating that a substantially larger population of children (compared to prior to the pandemic) will continue online school (Schwartz et al., 2020). This will include having social relationships with schoolmates and teachers that will be almost exclusively online. It already was a challenge for scientists studying school contexts to integrate traditional in-person settings with newer online school settings into their research. As the number of online schools grows, and as it becomes more common for schools to take blended "hybrid" approaches (in which some students attend in-person and some attend online, or all students attend partly in-person and partly online), it will be imperative for researchers to take these contexts into consideration in theorizing, hypothesis generation, sampling, and inferencing. Our conjecture based on theory is that there is no reason to hypothesize that the effects of negative peer or teacher relationship experiences on neurocognitive processes would be qualitatively different for in-person versus virtual contexts. Nevertheless, that speculation is based on no data—a gap in knowledge that we hope will begin to be filled by scholars and scientists in the years ahead.

Recommendation 2: Integrating Neurobiological Factors into Prevention and Intervention

Most of the empirical work reviewed in this book showed predictive associations between (specific forms of) elementary school adverse social experiences and neurobiological outcomes. And although these studies made a compelling case of studying social environmental factors and neurobiological factors in concert, this is just the first step. Future research should invest more in combining these domains in new and novel research designs. This may include assessing multiple neurobiological processes in concert, preferably nested within a cohort study with real-life measurements of elementary school social experiences (note also recommendation 3 below). This would enable studying the interplay between these neurobiological processes itself and their links with stable and dynamically changing aspects of children's social environmental experiences. A closely related avenue would be to incorporate neurobiological measures as outcomes of prevention efforts, to study if the detrimental processes that were described in this book can be reversed (e.g., epigenetics: Weaver et al., 2006; HPA axis function: Kuhlman et al., 2014).

Yet another approach would be to incorporate neurobiological correlates of detrimental social experiences as tools for promoting change within prevention and intervention efforts. For instance, biofeedback—combined with virtual reality tools—has been suggested as an approach for fostering human empathic skills (Schoeller et al., 2019) and to reduce classroom anxiety and disruptive behavior (Bossenbroek et al., 2020). Also, adding neurobiological measures in screening of children could be considered. Departing from the idea that the only effect of neurobiology is as a mediator in the causal pathway *from* adverse social experiences *to* negative outcomes, screening would involve measures on these potentially compromised neurobiological factors prior to exposure to adverse experiences and emergence of poor developmental outcomes. Doing so could powerfully augment the effective screening of children "at risk."

We must be concerned with ethical issues, especially if it comes to screening on neurobiological factors that identify young children whose underlying biological processes predispose them to evoking adverse peer and teacher social experiences once they begin school. The human research literature shows that the neurobiological factors outlined in the current volume are well established as *correlates* of adverse social environmental experiences, but we do not yet know the key causal pathways. In addition, we do not know how well neurobiological factors discriminate between children who are at higher versus lower risk of developing future negative outcomes. Humility regarding the limitations of our knowledge and caution regarding potential deleterious effects are essential when considering when and how to include neurobiology in prevention and intervention designs.

Recommendation 3: Embracing Diversity and Complexity of People and Systems

A major concern when testing complex systems is that there are criteria for determining scientific rigor that generally require parsimony. For example, anyone conducting experiments to test a hypothesis recognizes that they are testing only one mechanism (or even just one part of a mechanism) within a much larger complex system of mechanisms. Relatedly, the contemporary and growing emphasis on "open" science, reproducibility and replication, certainly promotes transparency and rigor (which is welcomed), but this challenges scientists who are striving to empirically test theories involving complex systems (Wentzel, 2021). This is a major challenge precisely because it is that much harder to replicate the methods (let alone findings) of a large-scope study that is simultaneously examining multiple variables, mechanisms, and pathways in a complex system. Nevertheless, theories of child development and the development of psychopathology all require a complex systems view, for hypothesis generation (Osher et al., 2020).

Our theories also require rigor with regard to addressing the remarkable diversity of people. Most notable in this regard is the long history of psychological and neuroscience researchers' emphasis on "WEIRD" (Western [and we would add, White], Educated, Industrialized, Rich, Democratic) samples that not only lack representation but that lead to findings that become instantiated as a "norm" (Henrich et al., 2010). We urge fellow scientists and scholars to take seriously the goal to do research that seeks to understand and optimize human development, particularly by focusing resources that promote research that supports minoritized scholars and research participants and that tackle inequities (Brown et al., 2019; for a recent example, see Wonkam, 2021, regarding the Three Million African Genomes [3MAG]).

One very common approach to the challenge of testing complexity is to vastly expand the multivariate measurement of constructs—that is, to include more methods for measuring more constructs in more parts of the system, with larger samples that are more representative (or that overrepresent previously understudied populations), and across more time points. The effort has been greatly facilitated by the advent and widespread adoption of measurement, explanatory, and predictive modeling using structural equation models and the like (Kline, 2016). Similarly, there has been an expansion in consideration of the variety of qualitatively distinct subgroups of children that are present, based on distinct effects linking predictors and developmental outcomes (i.e., latent mixture modeling; Berlin et al., 2014). However, there are shortcomings with the standard approaches for testing large sets of variables, with respect to replicability of effects and "chance" findings and with regard to prediction and drawing conclusions (e.g., Chin et al., 2020). Furthermore, most of the modeling that is done typically examines linear additive effects, at the cost of testing and distinguishing from each other the presence of more complex nonlinear and interactive effects. Nevertheless, it is feasible to estimate such effects, and doing so holds promise for increasing the clarity of our understanding of underlying complexity in processes (Belzak & Bauer, 2019).

A more recent innovation to address some of the limitations of embracing complexity is to develop algorithms that are modeled on available patterns in data—that is, that explore the large dataset and "learn" from it and thereby include high degrees of multivariate interactive and nonlinear effects. The machine or "deep" learning approach (Alpaydin, 2020) has become common in fields that rely heavily on engineering and computer science tools and methods (e.g., computational biology, neuroscience, genetics, etc.) but, to our knowledge, have not found traction (yet) in child developmental science (e.g., Turgeon & Lanovaz, 2020). Also, as highlighted in several of the chapters (e.g., in chapter "How Peers and Teachers Shape Elementary School Children's Academic and Socioemotional Development" with respect to incorporating genetic and epigenetic information into our models), developmental scientists will need to work with much larger samples of participants than we typically do, due to the statistical power requirements for such work. It will be essential as a field for us to embrace the complexity of our theories in our empirical approaches, if we are to make progress in integrating the myriad sources of essential information into the next generation of descriptive, explanatory, and individually predictive models.

Concluding Thoughts

The current book is not the first to show the terrible and sometimes horrific consequences of damaging social experiences during formal schooling, such as being the victim of bullying, becoming socially rejected by classmates, or experiencing poor support by teachers. The novelty instead rests in its overview of contemporary and future-looking research that examines how adverse social experiences already during the elementary school period may become embedded in the neurobiology of children as potential causal pathways in the well-documented link between harmful social experiences and long-term negative outcomes. The neurobiological processes the authors have examined and present capture likely (and at least possible) causal pathways that have all been associated with elementary school social experiences. However, as noted above, it is not likely that each of these neurobiological processes operates only as a mediator to outcomes; the biological "consequences" themselves play an important causal role in furthering subsequent adverse or ameliorative social experiences by shaping children's foci of attention and other aspects of processing and interpreting social events in ways that make it more difficult to improve their relationships with peers and teachers. Furthermore, each of the biological processes operates together in a complex system and not in isolation and can coalesce into a negative spiral that over time becomes increasingly difficult to modify.

It is yet largely unknown whether the biological scars that follow elementary school poor social experiences can be reversed. As noted above, some studies do suggest this is possible and provide important leads for future prevention and intervention. However, and notwithstanding the importance of these new leads, we must address the initial "cause" in the chain as something that we should not tolerate (regardless of its consequences)—the mistreatment of children by their peers and teachers in school settings. Almost all elementary school children and teachers, in some way or another, are part of the adverse social conditions described in this book: as the perpetrator, facilitator, witness, or victim. This should not be the case. We have noted that the current book is not the first regarding the detrimental effects of harmful peer and teacher social experiences on children's development. Unless policymakers, researchers, and practitioners work together to prioritize the elimination of adversity in school settings, the current book will not be the last on this topic either. We hope that in ways big and small, the knowledge and ideas put forward by the chapters' authors will contribute to improvements in children's daily lives and developmental outcomes.

References

Alpaydin, E. (2020). *Introduction to machine learning* (4th ed.). MIT Press.

Belzak, W. C., & Bauer, D. J. (2019). Interaction effects may actually be nonlinear effects in disguise: A review of the problem and potential solutions. *Addictive Behaviors, 94*, 99–108.

Berlin, K. S., Williams, N. A., & Parra, G. R. (2014). An introduction to latent variable mixture modeling (part 1): Overview and cross-sectional latent class and latent profile analyses. *Journal of Pediatric Psychology, 39*(2), 174–187.

Bossenbroek, R., Wols, A., Weerdmeester, J., Lichtwarck-Aschoff, A., Granic, I., & van Rooij, M. M. (2020). Efficacy of a virtual reality biofeedback game (DEEP) to reduce anxiety and disruptive classroom behavior: Single-case study. *JMIR Mental Health, 7*(3), e16066.

Brown, C. S., Mistry, R. S., & Yip, T. (2019). Moving from the margins to the mainstream: Equity and justice as key considerations for developmental science. *Child Development Perspectives, 13*(4), 235–240.

Chin, W., Cheah, J. H., Liu, Y., Ting, H., Lim, X. J., & Cham, T. H. (2020). Demystifying the role of causal-predictive modeling using partial least squares structural equation modeling in information systems research. *Industrial Management & Data Systems, 120*(12), 2161–2209.

Henrich, J., Heine, S. J., & Norenzayan, A. (2010). The weirdest people in the world? *Behavioral and Brain Sciences, 33*(2–3), 61–83.

Keith, S., & Martin, M. E. (2005). Cyber-bullying: Creating a culture of respect in a cyber world. *Reclaiming Children and Youth, 13*(4), 224–228.

Kline, R. B. (2016). *Principles and practice of structural equation modeling* (4th ed.). Guilford Press.

Kuhlman, K. R., Olson, S. L., & Lopez-Duran, N. L. (2014). Predicting developmental changes in internalizing symptoms: Examining the interplay between parenting and neuroendocrine stress reactivity. *Developmental Psychobiology, 56*(5), 908–923.

Osher, D., Cantor, P., Berg, J., Steyer, L., & Rose, T. (2020). Drivers of human development: How relationships and context shape learning and development. *Applied Developmental Science, 24*(1), 6–36.

Schoeller, F., Bertrand, P., Gerry, L. J., Jain, A., Horowitz, A. H., & Zenasni, F. (2019). Combining virtual reality and biofeedback to foster empathic abilities in humans. *Frontiers in Psychology, 9*, 2741.

Schwartz, H., Grant, D., Diliberti, M., Hunter, G., & Setodji, C. (2020). *Remote learning is here to stay: Results from the first American School District panel survey.* Rand Corporation. URL: https://www.rand.org/pubs/research_reports/RRA956-1.html. Creative commons attribution 4.0 international public license, RR-A956-1, 2020. Accessed 6 June 2021.

Turgeon, S., & Lanovaz, M. J. (2020). Tutorial: Applying machine learning in behavioral research. *Perspectives on Behavior Science, 43*(4), 697–723.

Twenge, J. M., Joiner, T. E., Rogers, M. L., & Martin, G. N. (2018). Increases in depressive symptoms, suicide-related outcomes, and suicide rates among US adolescents after 2010 and links to increased new media screen time. *Clinical Psychological Science, 6*(1), 3–17.

Weaver, I. C., Meaney, M. J., & Szyf, M. (2006). Maternal care effects on the hippocampal transcriptome and anxiety-mediated behaviors in the offspring that are reversible in adulthood. *Proceedings of the National Academy of Sciences, 103*(9), 3480–3485.

Wentzel, K. R. (2021). Open science reforms: Strengths, challenges, and future directions. *Educational Psychologist, 56*(2), 161–173.

Wonkam, A. (2021). Sequence three million genomes across Africa. *Nature, 590*, 209–211.

World Bank. (2021). How countries are using edtech to support access to remote learning during the Covid-19 pandemic. URL: https://www.worldbank.org/en/topic/edutech/brief/how-countries-are-using-edtech-to-support-remote-learning-during-the-covid-19-pandemic. Accessed 6 June 2021.

Index

P. A. C. van Lier, K. Deater-Deckard (eds.), *Biosocial Interplay During Elementary School*, https://doi.org/10.1007/978-3-031-07109-6

The manufacturer's authorised representative in the EU is Springer Nature Customer Service Centre GmbH, Europaplatz 3, 69115 Heidelberg, Germany. If you have any concerns regarding our products, please contact ProductSafety@springernature.com

Printed and bound by CPI Group (UK) Ltd, Croydon, CR0 4YY
22/09/2026
02221055-0003